Springer Tracts in Modern Physics
Volume 125

Editor: G. Höhler
Associate Editor: E. A. Niekisch

Springer Tracts in Modern Physics

Volumes 90–111 are listed on the back inside cover

* denotes a volume which contains a Classified Index starting from Volume 36

Eberhard Burkel

Inelastic Scattering

of X-Rays
with Very High Energy Resolution

With 70 Figures

Springer-Verlag
Berlin Heidelberg GmbH

Dr. habil. Eberhard Burkel

Sektion Physik der Ludwig-Maximilians-Universität München
Geschwister-Scholl-Platz 1, W-8000 München 22, Fed. Rep. of Germany

---

Manuscripts for publication should be addressed to:

Gerhard Höhler

Institut für Theoretische Kernphysik der Universität Karlsruhe, Postfach 69 80,
W-7500 Karlsruhe 1, Fed. Rep. of Germany

*Proofs and all correspondence concerning papers in the process of publication
should be addressed to*

Ernst A. Niekisch

Haubourdinstraße 6, W-5170 Jülich 1, Fed. Rep. of Germany

ISBN 978-3-662-15009-2

Library of Congress Cataloging-in-Publication Data. Burkel, Eberhard, 1952– . – Inelastic scattering of x-rays with very high energy resolution / – Eberhard Burkel. – p. cm. -- (Springer tracts in
modern physics ; v. 125). – Includes bibliographical references and index.
ISBN 978-3-662-15009-2   ISBN 978-3-540-38351-2 (eBook)
DOI 10.1007/978-3-540-38351-2
1. X-rays--Scattering. I. Title. II. Series: Springer tracts in modern physics;
125.QC1.S797 vol. 125   [QC482.S3]   530 s--dc20   [539.7'222]   91-31752

© Springer-Verlag Berlin Heidelberg 1991
Originally published by Springer-Verlag Berlin Heidelberg New York in 1991
Softcover reprint of the hardcover 1st edition 1991

Typesetting: Camera ready by author
57/3140-543210 – Printed on acid-free paper

Preface

X-rays are an established tool for investigations of condensed matter in solid state physics and crystallography. The standard analysis of the elastically scattered X-rays reveals the short and the long range order of atoms. It gives information on the structure of atoms as well as on their local displacements.

In these studies, the high intensity of the new generation of X-ray sources – synchrotrons and storage rings – has opened fantastic new research areas. In particular, scattering experiments with X-rays can be performed with resolutions in energy, space and time which were not thinkable before the advent of these sources.

In general, textbooks on solid state physics dealing with lattice dynamics come to the conclusion that lattice vibrations cannot be resolved directly with X-rays. About 100 years after the discovery of X-rays it has now become possible to achieve sufficiently high energy resolution for the direct observation of the small energy shifts of the photons due to the creation or annihilation of lattice vibrations in an inelastic scattering experiment.

The aim of this review is to report on the recent progress in the field of inelastic scattering of X-rays with very high energy resolution, which is directly correlated with the development of the spectrometer INELAX. This instrument uses the backscattering technique (Bragg angles close to 90°) of X-rays at the storage ring DORIS at HASYLAB, DESY in Hamburg, and has recently yielded an energy resolution of 9 meV, which corresponds to a relative energy resolution of about $5 \cdot 10^{-7}$.

After short historic comments and a comparison between different probes for inelastic investigations for condensed matter excitations, the basic principles of the backscattering technique are described. The technical realization of the spectrometer INELAX is discussed in detail. Similar attempts of instrumental set-ups undertaken in other laboratories are reported, too.

The second half of the review covers the first applications of the inelastic scattering technique. The selection of the different examples aimed for a broad spectrum of problems in order to demonstrate the flexibility and power of this rapidly growing research field.

The feasibility of the determination of phonon dispersion curves in single crystals is demonstrated on single crystals of beryllium and diamond. The results on the longitudinal and transverse modes from X-ray scattering are in excellent agreement with the results from thermal neutron scattering, the conventional method for measuring phonon dispersion curves.

The application of this new method to the field of biology is certainly a challenging aspect. The power of the technique is shown by the measurements of internal and external modes of vibrations in the crystallized amino acids alanine and glycine up to energy transfers of 500 meV.

Collective excitations in liquids will be a dominating research area for inelastic X-ray scattering. First observations of the dispersion of such excitations in liquid lithium are discussed as well.

The high energy resolution will certainly give new impulse to study the electronic excitations in solids and liquids, too. Measurements of such excitations in single crystals of lithium up to energy transfers of 5 eV, for instance, provide information on the dispersion of the so-called zone boundary collective states.

The final discussion considers the further applications of inelastic X-ray scattering that would become possible with continued technological improvements.

Munich, August 1991   *Eberhard Burkel*

Acknowlegements

The author thanks J. Peisl and B. Dorner for fruitful discussions and encouraging support and all members of the INELAX team for their productive collaboration over the years, especially Th. Illini for his enthusiasm in completing the computer software.

HASYLAB provided hospitality and technical assistance, which is highly appreciated.

The support and hospitality of M. Bartunik and his group at the Max-Planck Institute at DESY is appreciated.

The excellent collaboration with S. Lederle in the construction of instrument components was very productive. The instrument was built with the technical support of A. Ebenböck and his crew at the machine shop of the Sektion Physik of the University of Munich. Special thanks to them and to all who contributed to the success of INELAX.

The author thanks S. Motz for drawing the figures of this review and he greatly appreciates the careful reading of the manuscript by G. Materlik, W. Schülke, H. Sinn and by A. M. Köhler whose encouragement is especially appreciated.

This project is supported by the Bundesministerium für Forschung und Technologie under the project number 03 PE1 LMU 2.

Contents

Abbreviations and Symbols

a_0	lattice parameter
A	mismatch factor; (5.1)
A	Thomson part of scattering amplitude; (7.5)
b	scattering length of nucleus
B	bulk modulus
BCS	Bardeen-Cooper- Schriefer theory
c	velocity of light
c_0	calibration constant; (4.2)
$c_1, \ldots, c_4$	constants; (4.3)
c_{11}, c_{44}	elastic constants
C	prefactor in scattering amplitude; (7.6)
constant-Q scan	scan with fixed momentum transfer
constant-E scan	scan with fixed energy transfer
d, d'	distance between two beams
d_{hkl}	lattice spacing for $(h\ k\ l)$ reflection
D	diameter of spherical crystal
$D_{\mathrm{hor}}, D_{\mathrm{ver}}$	horizontal and vertical beam sizes at monochromator
$D'_{\mathrm{hor}}, D'_{\mathrm{ver}}$	horizontal and vertical beam sizes at analyzer
DESY	Deutsches Elektronen Synchrotron
DORIS	electron-positron storage ring at DESY, Hamburg
$d\Omega$	solid angle element
$(d\sigma/d\Omega)_0$	intrinsic scattering cross section
$(d\sigma/d\Omega)_{\mathrm{Th}}$	Thomson scattering cross section
$(d\sigma/d\Omega)_{\mathrm{Ruth}}$	Rutherford scattering cross section
$d^2\sigma/(d\Omega d\omega_{\mathrm{f}})$	double differential scattering cross section; (2.1)
e	charge of electron
eV	electron Volt
$e_d(q, j)$	normalized phonon eigenvector of mode j with phonon wavevector q for atom d
$e_{\mathrm{i}}, e_{\mathrm{f}}$	initial and final polarization of the photon beam
E	energy transfer in the scattering process; (2.2)

E_c	critical energy
E_i, E_f	initial and final energy of photons
E_{el}	energy of the electron beam
EELS	electron energy loss spectroscopy
ESRF	European Synchrotron Radiation Facility, Grenoble
$f(Q), f_d(Q)$	atomic form factor, general and of the atom d
$f'(E_i), f''(E_i)$	contribution to atomic structure factor
$F(Q)$	structure factor
$F(\omega, T, q, j)$	function given by (6.3)
G	rigidity modulus
GeV	Giga electron Volt $= 10^9$ eV
$G(Q, q, j)$	dynamical structure factor for one phonon scattering; (6.2)
h_s, h_{hor}, h_{ver}	source size, its horizontal and vertical component
$\hbar$	Planck's constant
$\hbar Q$	momentum transfer in the scattering process; (2.2)
$\hbar Q$	absolute value of the momentum transfer; (2.3)
$(h\ k\ l)$	Miller indices
HARWI	Hard X-ray wiggler at the DORIS storage ring
HASYLAB	Hamburger Synchrotronstrahlungs Labor
ILL	Institute Max von Laue - Paul Langevin, Grenoble
$Im[-1/\varepsilon(Q, \omega)]$	energy loss function
INELAX	instrument for inelastic X-ray scattering
$I(Q, \omega)$	measured scattering intensity
k_B	Boltzmann's constant
k_F	Fermi wavevector
k	wavevector of the photon
k, k_0	absolute values of the wavevectors $k = 2\pi/\lambda$
k_{max}, k_{min}	absolute maximum and minimum value of the wavevectors
k_i, k_f	initial and final wavevector of the photon
keV	kilo electron Volt $= 10^3$ eV
$K_{\alpha_{1,2}}$	characteristic X-ray wavelength
l, l', L, L'	geometrical distances
LA, LO	longitudinal acoustic and optic phonon polarization
m	mass of electron or particle
M_j	scattering amplitude of electron j; (7.4)
M_d	mass of atom d
meV	milli electron Volt $= 10^{-3}$ eV

$\langle n \rangle$	Bose occupation factor
N	number of unit cells in crystal or number of particles particles
neV	nano electron Volt $= 10^{-9}$ eV
PEP	electron-positron storage ring at Stanford Synchrotron Laboratory
PETRA	high energy ring at DESY, Hamburg
$\boldsymbol{Q}, Q$	scattering vector and its absolute value
$\boldsymbol{Q}_0$	wavevector of the first maximum of the liquid structure factor
$\boldsymbol{Q}_\mathrm{c}$	cutoff wavevector of the plasmon
$(\boldsymbol{Q}, \omega)$	momentum and energy transfer space
SSRL	Stanford Synchrotron Radiation Laboratory
T, T_0	temperatures
$T_\mathrm{ana}, T_\mathrm{mono}$	temperature of analyzer and monochromator
TA, TO	transverse acoustic and optic phonon polarization
t_E	extinction length
r_0	classical electron radius
r_B	Bohr radius
$\boldsymbol{r}_j(t)$	vector to the position of particle j
R	bending radius
R_0	hard sphere radius
$R(\boldsymbol{Q}, \omega)$	resolution function
$S(\boldsymbol{Q})$	static structure factor
$S(\boldsymbol{Q}, \omega)$	scattering function; (2.5) – (2.8)
v	velocity of sound or of particles
V	volume of the unit cell
$w_\mathrm{hor}, h_\mathrm{ver}$	horizontal and vertical sizes of crystal elements
W, W_d	exponent in the Debye-Waller factor
W2	HARWI wiggler at DORIS
Z	atomic number
ZBCS	zone boundary collective state
$\alpha(T)$	thermal expansion coefficient
β	$\beta = 1 / k_\mathrm{B} T$
γ	reduced electron energy E_el / mc^2

Γ parameter describing crystal thickness

δE absolute energy resolution
$\delta E/E$ relative energy resolution
δk uncertainty of the wavevector k
$\delta k/k$ relative variation of the wavevector
$(\delta k/k)_\tau$ crystal contribution to the relative variation of the wavevector
$(\delta k/k)_\varepsilon$ contribution of the divergence to the relative variation
            of the wavevector
δt time uncertainty
$\delta(x)$ Dirac delta function; (2.5) and (6.3)
$\delta\beta$ beam divergence
$\delta\varepsilon, \delta\tilde{\varepsilon}$ angular divergence of the beam
$\delta\tau$ extension of the reciprocal lattice point
$\delta\psi_{\mathrm{el}}$ vertical divergence of the electron beam
$\delta\psi_{\mathrm{R}}$ energy dependent divergence of the synchrotron beam
$\delta\psi_{\mathrm{tot}}$ total vertical divergence of the synchrotron radiation beam

Δt time pulse
ΔT temperature difference

$\varepsilon, \varepsilon_0$ angular deviation from backscattering
$\varepsilon(\boldsymbol{Q}, \omega)$ dielectric function

ζ_{char} characteristic or periodic length

θ scattering angle

λ wavelength
λ_{c} characteristic wavelength
$\lambda_{\mathrm{K}\alpha_{1,2}}$ wavelength of characteristic X-ray line
μ linear absorption cofficient
$\mu\mathrm{eV}$ 10^{-6} eV

ρ_0 density

σ_j spin of electron j

ς oscillator damping constant

τ reciprocal lattice vector

τ_{r} relaxation time

φ angle

χ_s	adiabatic compressibility
$\chi(Q,\omega)$	dynamic susceptibility
ψ	elevation angle from electron orbit
ω	frequency
ω_e	frequency of Einstein oscillator
ω_i, ω_f	initial and final frequency
ω_{Gap}	gap frequency
ω_r	resonance frequency
$\omega_{q,j}$	frequency function of the phonon mode j with wavevector q
ω_0^2	normalized second moment of the particle response function
ω_1^2	normalized fourth moment of the particle response function
$|\Upsilon_i\rangle, |\Upsilon_f\rangle$	initial and final quantum states

1. Introduction

1.1 X-ray Diffraction

Since its discovery (Friedrich, Knipping and Laue 1912; Bragg 1913), X-ray diffraction has been an important and successful source of information on the local arrangement of atoms and molecules in condensed matter.

In the early days of X-ray diffraction, scientists were seriously discussing whether the strong thermal movements of the atoms in a crystal would allow the observation of diffraction spots at all. The experiments soon proved that they were observable. The thermal movements of atoms were found to be responsible for an attenuation of the intensities of the Bragg peaks by a Debye-Waller factor (Debye 1913, 1914; Schrödinger 1914; Faxén 1918, 1923; Waller 1923) and for the appearance of diffuse scattering intensity between the peaks. It was concluded that the thermal diffuse scattering intensity contained information on the lattice dynamics, and attempts on the interpretation were made. For a detailed historic overview see Ewald (1962).

Laval (1938, 1939) and James (1948) related the diffuse scattering to the frequency and wavelength of traveling elastic waves in the crystal and described the thermal motion of the lattice by a linear superposition of these waves.

From measurements of the diffuse scattering intensity along high symmetry directions in reciprocal space, dispersion curves for waves propagating along these axes could be derived. In combination with the theory of lattice dynamics by Born (1942), the tensor of the interatomic force constants could be derived. Curien (1952) and Jacobsen (1955) determined general force constants for first-, second- and third-nearest neighbor interactions for α-iron and copper. Joynson (1954) used a similar method for the hexagonal lattice of zinc and Walker (1956) for fcc aluminum.

Thermal diffuse scattering is not an elastic process, because the energy of a photon can be changed by the creation or annihilation of vibrational quanta, the phonons; thus, direct measurement of these energy shifts could provide information on the lattice vibrations in materials. Yet, X-rays with wavelengths comparable to interatomic distances have energies (10 keV) of about six orders of magnitude higher than typical phonon energies. Therefore, it is rather difficult to measure the energy change of the scattered photons directly.

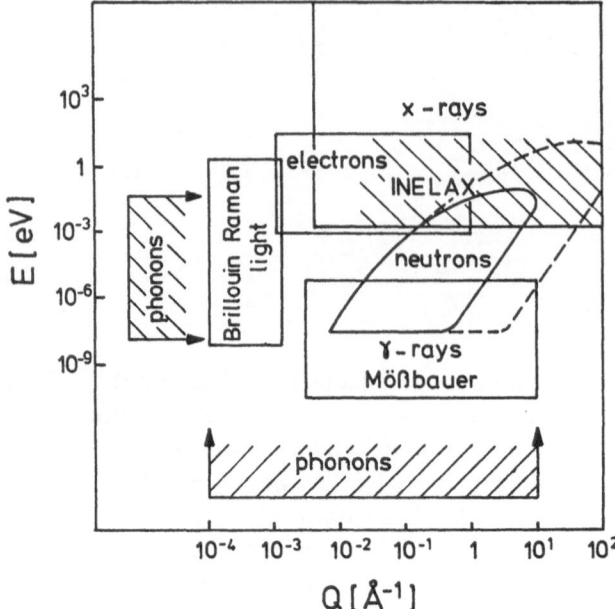

Fig. 1.1. Accessible ranges in the energy-momentum transfer space for the different probes of inelastic scattering

Only thermal energy neutrons allowed direct measurements of the small energy transfers associated with lattice vibrations. Therefore, inelastic neutron scattering became the leading method for determining the energies of phonon excitations (Brockhouse, Arase, Cagglioti, Rao and Woods 1962; Bacon 1962).

Only the advent of high intensity X-rays emitted by synchrotrons and storage rings made the direct measurement of phonons with inelastic scattering of X-rays feasible (Burkel, Peisl and Dorner 1987).

1.2 Probes for Inelastic Scattering with High Energy Resolution

Low energy excitations in solids and liquids can be investigated at present with several excellent energy resolving experimental methods. Figure 1.1 demonstrates the utilization of the different methods by showing schematically the range of energy and momentum transfers. Experiments which utilize inelastic scattering of neutrons, including spin echo methods, cover a wide range of energy resolution δE, up to the value $\delta E = 10^{-8}$ eV for those excitations having a detectable neutron scattering cross section. The increase of the accessible range expected by the new generation of neutron spallation sources is also indicated by dashed lines in Fig. 1.1.

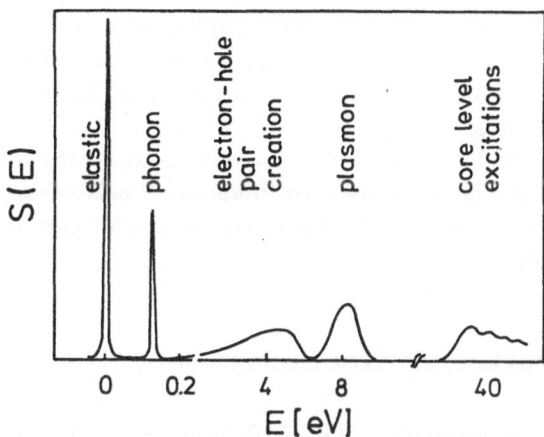

Fig. 1.2. Typical X-ray excitation profile in condensed matter, drawn schematically as a function of the excitation energy. It shows contributions due to elastic scattering and phonon scattering. Electron-hole pair creation, plasmonic and core level excitations are indicated as well

Ultrasonic measurements can be used to determine the elastic moduli of crystals with sufficiently good resolution. Light scattering in the Raman and Brillouin regime is an established method for optically active excitations. These high resolution experiments extend in energy resolution to $\delta E = 10^{-8}$ eV, but only for momentum transfers close to zero. Spectroscopy using the Mößbauer effect, recently combined with synchrotron radiation by nuclear Bragg diffraction experiments (Gerdau, Winkler, Tolksdorf, Klages and Hannon 1985), has an energy resolution as high as $\delta E = 10^{-10}$ eV; however, these measurements are limited to excitations close to zero energy transfer ($E < 5 \cdot 10^{-5}$ eV). Electron scattering covers a regime in energy-momentum space which is partly overlapping with X-rays and neutrons and is also indicated in Fig. 1.1.

Energy and momentum transfers, correlated with the investigations of phonons, are shown in Fig. 1.1, thus demonstrating the applicability of the different methods for their detection.

A typical X-ray excitation spectrum for condensed matter is shown schematically in Fig. 1.2. The elastic line at zero energy transfer is indicated as well. The lattice and molecular vibrations have typical energies up to several hundred meV. In the low-eV regime electron-hole pair creation is detectable. Plasmon excitations are correlated with energies around 5 to 20 eV. Core shell excitations require drastically higher energy transfers.

For the investigations of the described excitations with inelastic X-ray scattering appropriate energy resolution is very important. A resolution of about 10 meV is necessary for the study of vibrational excitations.

The first X-ray experiments for the analysis of electronic excitations were performed with limited energy resolution (for an overview see Platzman 1974).

A new generation of experiments started when the strong synchrotron sources allowed an energy resolution of about 1 eV and a high signal-to-noise ratio (Schülke et al. 1986, 1987). An improved energy resolution of about 100 meV will further stimulate this expanding research field.

The range of 10 to 100 meV resolution can be achieved in X-ray scattering by using backscattering geometry. Excitations with momentum and energy transfers lying in the shaded range of Fig. 1.1 (marked INELAX) can be directly studied with this technique.

1.3 The Use of Backscattering Geometry

Already in the early days of X-ray diffraction a lot of effort was spent for the precise determination of lattice parameter and elastic stresses. Pioneer work of van Arkel (1927) demonstrated that the precision of such experiments can be improved by using Bragg reflections close to 90° and by extrapolating the lattice parameter to the angle of 90° in order to reduce the influence of errors. Precise measurements in close to backscattering geometry were performed by Sachs and Weert (1930). That work was part of the basis for the technological applications of stress studies in materials, lateron.

An improvement of the resolution for the determination of the lattice parameter was proposed by Bottom (1965) with a scattering experiment in a backreflection geometry. The first experimental study of the proposed arrangement was performed with neutrons at the Munich reactor by Alefeld in 1966. Two major applications of the method were demonstrated at that time, which were the precise determination of lattice parameter changes and the possibility of inelastic scattering with very high energy resolution. The backscattering technique was further developed for neutron inelastic scattering by Alefeld, Birr and Heidemann (1968) and Heidemann (1970). This method is used as a standard tool in the neutron research field today.

Backscattering of X-rays based on Bottom's idea was first performed by Sykora and Peisl in 1970, and later by several other groups (Freund 1971; Bottom and Carvalho 1971; Freund and Schneider 1972; Okasaki and Kawami-nami 1973; Peisl 1976). These experiments used elastic scattering techniques to determine lattice parameter changes and lattice strains.

The use of this method for inelastic X-ray scattering failed at that time, because the intensity even from an X-ray generator with a rotating anode was considered to be too low for successful experiments (Dorner and Comès 1977). Since no characteristic line was compatible with the lattice spacing of the monochromator at room temperature in backscattering geometry, the wavelength had to be selected from the bremsstrahlung continuum, thus offering insufficient intensity.

With the development of storage rings, sources of white X-ray radiation, the situation changed. Several groups started projects in Hamburg (Graeff and

Materlik 1982), Brookhaven (Fuji, Hastings, Ulc and Moncton 1982; Moncton and Brown 1983) and Munich (Dorner and Peisl 1983; Benda, Dorner and Peisl 1983). At the same time, a project study at an x-ray generator was started in Bayreuth (Egger, Hofmann and Kalus 1984) with an expected energy resolution between 5 and 40 meV. Improvements of spectrometers using Fabry-Perot type interferometers for the monochromator and the analyzer (Steyerl and Steinhauser 1979) have been proposed for obtaining a resolution of $\delta E = 0.40$ meV. For an overview on the described activities see also Dorner (1984).

The first successful feasibility study of X-ray backscattering close to 90° at a storage ring was performed by Graeff and Materlik (1982). They achieved an energy width of $\delta E = 8$ meV with a simple two crystal arrangement without a scattering sample. Recently, the Bayreuth group (Egger, Hofmann and Kalus 1984; Hofmann 1989) achieved a resolution of $\delta E = 42$ meV with an experimental set-up using X-rays from a rotating anode by heating the monochromator and analyser crystals, which allowed the use of a characteristic line. The Brookhaven group achieved the same resolution in a test experiment at the PEP undulator (Moncton, Hastings, Siddons and Brown 1986) (see Chap. 5).

Our group has been developing an instrument for inelastic X-ray scattering at the DORIS storage ring of HASYLAB at DESY, Hamburg, since 1983 (Dorner and Peisl 1983; Benda, Dorner and Peisl 1983; Dorner, Benda, Burkel, Peisl 1985; Dorner, Burkel and Peisl 1986). The breakthrough with this instrument occurred in 1986, when the first signal from inelastic scattering by phonons was observed (Burkel, Peisl and Dorner 1987).

It is the aim of this review to discuss the technique and the performance of the inelastic X-ray backscattering instrument INELAX and to give an overview of the first applications to different research topics.

2. Basic Considerations

2.1 Scattering Cross Section

The results of a scattering experiment are usually compared with theory by means of the scattering cross section.

A general scattering experiment is shown schematically in Fig. 2.1. This arrangement is valid for all probes such as neutrons, electron beams and electromagnetic radiation, which were discussed in Chap. 1. The incident beam of well defined wavevector k_i, energy E_i and polarization unit vector e_i is scattered into the solid angle element $d\Omega$ under the scattering angle 2θ. The scattered beam is completely defined by the new wavevector k_f, the energy E_f and the polarization unit vector e_f.

The scattered intensity is described with the help of the double differential cross section,

$$\frac{d^2\sigma}{d\Omega \, d\omega_f} \, , \tag{2.1}$$

which decouples the specific quantities of the experiment (see below). It is given by the removal rate of particles out of the incident beam as the result of being scattered by N scatterers into a solid angle $d\Omega$ with a frequency range of $d\omega_f$.

The scattered beam is usually distributed over a range of energies E_f. There can be contributions of the beam that have been scattered elastically with no change of energy and other contributions that have changed energy due to inelastic scattering. Therefore, the scattering process contains information on energy and momentum transfers by

$$E \;=\; \hbar\omega \equiv E_i - E_f \quad \text{and}$$
$$\hbar Q \;\equiv\; \hbar \left(k_i - k_f \right) . \tag{2.2}$$

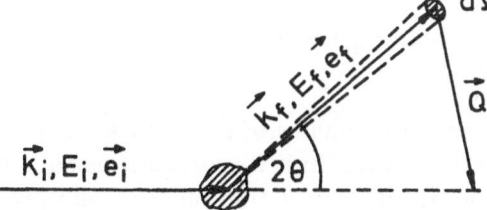

Fig. 2.1. Scattering geometry

As long as $E \ll E_i$ holds for the transferred energy E the momentum transfer $\hbar Q$ is simply connected with the scattering angle θ by

$$\hbar Q = 2\hbar \cdot k_i \cdot \sin\theta . \tag{2.3}$$

If there is a weak coupling between the probe and the target, the scattering can be treated in the lowest order Born approximation.

For the further discussion of the double differential scattering cross section the very elegant and informative presentation,

$$\frac{d^2\sigma}{d\Omega\, d\omega_f} = \left(\frac{d\sigma}{d\Omega}\right)_0 \cdot S(\boldsymbol{Q},\omega) , \tag{2.4}$$

derived by Van Hove (1954), will be used. This description has the advantage of allowing us to separate the double differential cross section into two contributions. The coupling of the beam to the scattering system is characterized by the intrinsic cross section $(d\sigma/d\Omega)_0$ and the properties of the sample in the absence of the perturbing probe are expressed by the scattering function $S(\boldsymbol{Q},\omega)$.

2.1.1 Scattering Function

The scattering function can be discussed in different modifications. A basic representation for a many-body system of particles at the positions $\boldsymbol{r}_j$ with discrete initial and final states Υ_i and Υ_f is expressed according to Fermi´s "golden rule" by

$$S(\boldsymbol{Q},\omega) = \sum_{\Upsilon_i,\Upsilon_f} \left| \left\langle \Upsilon_f \left| \sum_j e^{i\boldsymbol{Q}\boldsymbol{r}_j} \right| \Upsilon_i \right\rangle \right|^2 \cdot \delta\left(\hbar\omega + E_{\Upsilon_f} - E_{\Upsilon_i}\right) . \tag{2.5}$$

The phase of the scattering amplitude is described by $e^{i\boldsymbol{Q}\boldsymbol{r}_j}$. The matrix elements contain the probabilities for excitations from the initial to the final states and the delta function gives the corresponding frequency information. According to (2.5), the scattering amplitudes from the different scatterers are added and then squared. This leads to an interference between the scattering amplitudes.

A more sophisticated representation of the scattering function was derived by Van Hove (1954),

$$S(\boldsymbol{Q},\omega) = \frac{1}{2\pi} \int dt\, e^{-i\omega t} \left\langle \Upsilon_i \left| \sum_{j,j'} e^{-i\boldsymbol{Q}\boldsymbol{r}_j(t)} e^{i\boldsymbol{Q}\boldsymbol{r}_j(0)} \right| \Upsilon_i \right\rangle . \tag{2.6}$$

It describes the correlations of the scattering phases of the particles at positions $\boldsymbol{r}_j(t)$ at different times t. In the classical limit it represents essentially the Fourier transform in time of the density correlation function and gives information on the particle fluctuations in the scattering system in the same initial and final states at different times.

The sizes of the scattering phases are defined by the term $Q r_j$, which suggests to classify the scattering process by the magnitude of this term (Platzman 1974). The scattering system itself can be described by a characteristic length ζ_{char} which, for instance, can be an interparticle distance or a screening length. This length can be compared with the inverse of the wavevector Q as a measure for the transferred momentum $\hbar Q$.

For $Q \zeta_{\mathrm{char}} \ll 1$, there exists interference between the scattering amplitudes from many particles of the system. Consequently, mainly the collective behavior of the particles will be detectable. Therefore, collective motions of the scattering system like phonons, magnons or plasmons can be observed if, in addition, the transferred energy is in the characteristic frequency range of these motions.

For $Q \zeta_{\mathrm{char}} \gg 1$, the interference of the scattering amplitudes is negligible and the scattering contributions of different particles are independent. Therefore, single particle properties are observed, like, for example, Compton scattering (Platzman 1974) in the case of photon interaction with an electron system, if the photon energy is large compared to the binding energy of the electron.

In the intermediate ranges $Q \zeta_{\mathrm{char}} \approx 1$, both collective and single particle properties are visible.

Table 2.1 shows characteristic properties for X-rays, laser light, neutron and electron probes. Representative values for energies and wavelengths and the relations between them are given as additional information to Fig. 1.1. Typical values of the wavevector Q are shown as well. According to this, light scattering is sensitive only to long wavelength excitations and averages over thousands of atoms, whereas electron, neutron and X-ray scattering can resolve the atomistic structure.

An interesting comparison between the probes is to focus on the energy uncertainty δE_{i} of the scattering beam. It is correlated with the finite temporal extension δt of the incident waves by $\delta t \sim \hbar / \delta E_{\mathrm{i}}$. Table 2.1 reveals that X-rays, with $\delta t \sim 10^{-16}$ s, normally provide a "snapshot" picture of the atomic configuration, compared to other probes. X-rays are scattered at the instantaneous configuration since the movements of the atoms are slow (10^{-14} s) compared to the observation time. However, with the increased energy resolution the time scale of X-rays also allows to observe atomic motions.

The scattering function (2.6) is often transformed to representations that are more suitable to describe important physical properties of a particular system.

For the study of collective excitations, the response of a system initiated by the scattering photon, can also be expressed by the imaginary part or, in other words, the dissipative part of the dynamic susceptibility $\chi(Q, \omega)$ (Martin 1968):

Table 2.1. Typical energy values, relations between wavelength and energy, wavelengths and wavevectors for selected probes for inelastic scattering. Typical energy resolutions and correlated times are given as well. The marked values (*) are valid for X-ray scattering of high energy resolution with the instrument INELAX

Probe	Typical energy $E_i[eV]$	Conversion $\lambda[\text{Å}] \leftrightarrow E[eV]$	Wavelength $\lambda[\text{Å}]$	Wavevector $Q[\text{Å}^{-1}]$	Energy resolution $\delta E/E_i$	Time δt [s]
X-rays	10^4	$12400 \cdot E_i^{-1}$	1	10^{-2}–10	10^{-4}	10^{-16}
X-rays (*)					10^{-6}	10^{-14}
Laser light	1	$12400 \cdot E_i^{-1}$	$5 \cdot 10^3$	10^{-4}–10^{-3}	10^{-8}	10^{-8}
Neutrons	10^{-2}	$0.286 \cdot E_i^{-1/2}$	1	10^{-2}–10	10^{-4}	10^{-10}
Electrons	10^2	$12 \cdot E_i^{-1/2}$	1	10^{-3}–1	10^{-5}	10^{-13}

$$S(\boldsymbol{Q},\omega) = -\frac{1}{\pi} \cdot \frac{1}{1 - e^{-\beta\hbar\omega}} \cdot \text{Im}\,\chi(\boldsymbol{Q},\omega) \qquad (2.7)$$

with $\beta = 1/k_\text{B}T$.

For the special application to an electron gas with density N, the response function $\varepsilon(\boldsymbol{Q},\omega)$, called the dielectric function, is usually used to describe the scattering function (Pines and Noziéres 1966):

$$S(\boldsymbol{Q},\omega) = \frac{\hbar Q^2}{4\pi^2 e^2 N} \cdot \frac{1}{1 - e^{-\beta\hbar\omega}} \cdot \text{Im}\left[\frac{-1}{\varepsilon(\boldsymbol{Q},\omega)}\right] , \qquad (2.8)$$

where $\text{Im}\,[\,-1/\varepsilon(\boldsymbol{Q},\omega)]$ is the macroscopic energy loss function.

2.1.2 Intrinsic Scattering Cross Section for Different Probes

The coupling of the electromagnetic field of the photon beam to the scattering electron system of the sample is taken into account by the Thomson scattering cross section

$$\left(\frac{d\sigma}{d\Omega}\right)_0 = \left(\frac{d\sigma}{d\Omega}\right)_\text{Th} = r_0^2 \cdot (\boldsymbol{e}_i \cdot \boldsymbol{e}_f)^2 \cdot \left(\frac{\omega_f}{\omega_i}\right) \qquad (2.9)$$

of a single electron. $r_0 = e^2/mc^2 = 2.818 \cdot 10^{-13}$ cm is the classical electron radius. Therefore $(d\sigma/d\Omega)_\text{Th}$ is of the order of 10^{-25} cm^2.

The factor ω_f/ω_i in (2.9) is about equal to 1 in the normal case of X-ray scattering, because $E \ll E_i$, as mentioned before.

If visible light is used as the scattering probe, then the main coupling is due to the polarizability of the scattering system and the intrinsic cross section is

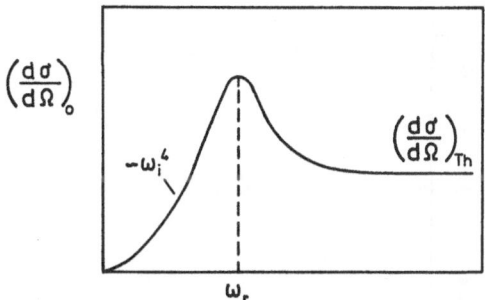

Fig. 2.2. Schematic view of the intrinsic cross section $(d\sigma/d\Omega)_0$ for the scattering of electromagnetic radiation in the light and X-ray range. ω_r is a typical resonance frequency of the scattering system and $(d\sigma/d\Omega)_{Th}$ is the Thomson scattering cross section

$$\left(\frac{d\sigma}{d\Omega}\right)_0 = r_0^2 \cdot (\mathbf{e}_i \cdot \mathbf{e}_f)^2 \cdot \frac{\omega_i^4}{\left(\omega_r^2 - \omega_i^2\right)^2 + \left(\varsigma\omega_i\right)^2} \,, \tag{2.10}$$

ω_r is a typical resonance frequency and ς is the corresponding oscillator damping constant. Therefore, light scattering is strongly frequency dependent (see Fig. 2.2). In the limit of light scattering at free electrons the Thomson cross section is obtained. Light scattering is sensitive to those collective excitations, which are correlated with a change of the polarizability. Scattering experiments performed with incident energies E_i close to $\hbar\omega_r$ are the class of resonance experiments which allow the enhancement of the scattering response.

Neutrons as probe in the scattering process are sensitive to the nuclei in the system. Only magnetic materials with appreciable electronic scattering are an exception. The nuclear scattering cross section can be written as

$$\left(\frac{d\sigma}{d\Omega}\right)_0 = \frac{k_f}{k_i} \cdot b^2 \,, \tag{2.11}$$

where b is the scattering length of the nucleus and its magnitude is of the order of 10^{-12} cm. Therefore, $(d\sigma/d\Omega)_0$ is of the order of 10^{-24} cm^2. The scattering length of nonmagnetic materials for neutrons is independent of Q due to the small dimensions of the nuclei ($\sim 10^{-12}$cm) relative to the used neutron wavelength. However, for X-ray scattering there is always a Q dependence due to the spatial distribution of the shell electrons of an atom. This is described by the atomic form factor $f(Q)$. It is determined by the Fourier transform of the electronic charge density distribution with $f(Q \to 0) = Zr_0$.

For neutron scattering, a similar Q dependence occurs for the magnetic scattering where the magnetic moment caused by unpaired electrons couples to the magnetic moment of the neutron. Normally, only a few electron orbits in the outer shell of the atom will contribute to the magnetic moment. Therefore, the magnetic scattering form factor for neutrons is not identical to the electronic form factor for X-rays.

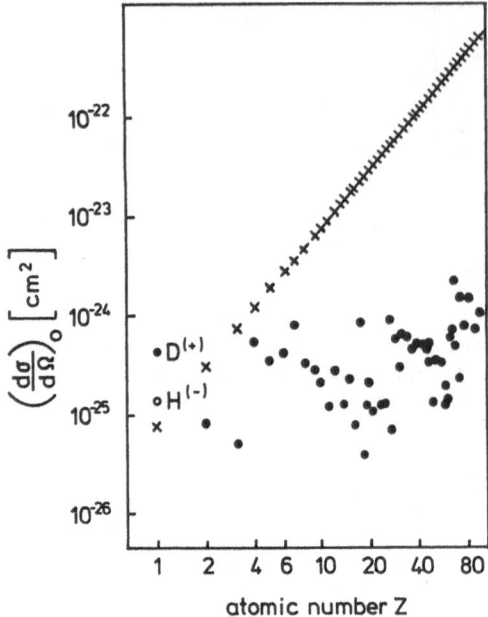

Fig. 2.3. Intrinsic cross sections for coherent scattering of neutrons ($\bullet$, $\circ$) and for X-ray scattering ($\times$) from atoms of the atomic number Z, in the forward direction ($Q \to 0$)

A comparison of the scattering cross sections of neutrons and X-rays is given in Fig. 2.3 by showing the intrinsic cross sections $(d\sigma/d\Omega)_0 = b^2$ for coherent nonmagnetic neutron scattering and $(d\sigma/d\Omega)_0 = Z^2 r_0^2$ for X-ray scattering from atoms of the atomic number Z in the forward direction ($Q \to 0$).

For further information on the behavior of the scattering lengths, on the influence of nuclear spin and isotopic incoherence and the different techniques of coherent and incoherent neutron scattering, the reader is referred to the variety of available textbooks on neutron scattering.

Inelastic excitations can also be studied with electron energy loss spectroscopy (EELS). The interactions of the high energy electrons with the scattering systems are described by the Rutherford scattering cross section (Zacharias 1975),

$$\left(\frac{d\sigma}{d\Omega}\right)_0 = \left(\frac{d\sigma}{d\Omega}\right)_{\text{Ruth}} = \frac{4}{r_B^2 Q^4} \, , \tag{2.12}$$

with $r_B = \hbar^2/me^2 = 5.292 \cdot 10^{-9}$ cm being the Bohr radius. For a scattering vector of $Q = 1$ Å^{-1}, for instance, $(d\sigma/d\Omega)_{\text{Ruth}}$ is of the order of 10^{-15} cm^2. This reveals a coupling in the scattering process several orders of magnitude higher than for the scattering of X-rays or neutrons. The consequence is the appearance of strong multiple scattering processes and extinction. Therefore, measuring the transmitted electron intensity is possible only when scattering through thin films. It will also show strong surface scattering effects (see also Sect. 6.4).

Table 2.2. Typical intrinsic atomic cross sections for different probes. r_0 is the classical electron radius, r_B the Bohr radius, b the scattering length and Z the atomic number

Probe	Intrinsic cross section	Magnitude [cm^2]
X-rays	$(Z\,r_0)^2$	10^{-25}
Electrons	$(2Z\,/\,r_B\,Q^2)^2$	10^{-15}
Neutrons	b^2	10^{-24}

Other aspects of X-ray scattering such as the anomalous and magnetic scattering, will be briefly discussed in Sect. 7.1 and a more detailed double differential scattering cross section can be found there.

2.2 Instrumental Principle

The basic idea for an inelastic scattering instrument shown in Fig. 2.4 is the same for X-ray and for neutron scattering. It is the general scheme of a three axis spectrometer. The beam, containing neutrons or photons, is well defined in energy E_i and wavevector k_i by elastic scattering from a monochromator crystal. After the inelastic scattering process at the sample the beam is analyzed for energy E_f and wavevector k_f, by elastic scattering at the analyzer crystal. This defines the momentum transfer and the energy transfer in the scattering process at the sample, according to (2.2).

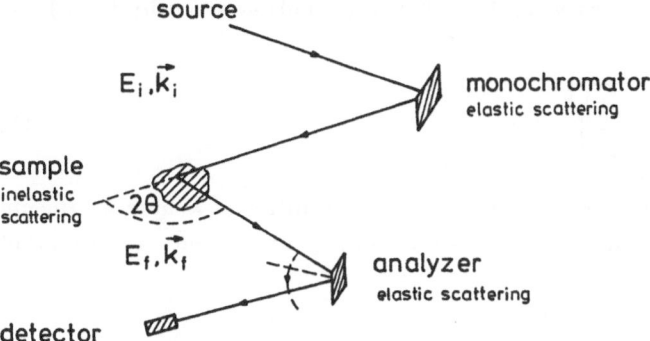

Fig. 2.4. Schematic of a three axis spectrometer with the main components for inelastic scattering

The variation of the amount of energy transferred can be achieved either by varying the scattering angles at the analyzer crystal or by varying the lattice parameter of the analyzer crystal. The latter method implies a temperature variation of the crystal and is well suited for inelastic X-ray scattering. A scan with constant scattering angle keeps the momentum transfer fixed and represents a so called constant-Q scan (Lovesey and Springer 1977).

Such an instrument can be used for inelastic measurements with X-rays only if its energy resolution is "good enough". Since "energy resolution" is the key word here, the different contributions to it will be discussed in some detail.

2.3 Energy Resolution

The relation of energy to wavevector for X-rays leads to a simple expression for the energy resolution

$$\frac{\delta E}{E} = \frac{\delta k}{k} \, . \tag{2.13}$$

If a beam with the wavevector k is scattered from a crystal with the Bragg angle θ, Bragg's law can be written as

$$2 \cdot k \cdot \sin \theta = \tau \quad \text{or} \quad k = \frac{\tau}{2 \cos \varepsilon} \, , \tag{2.14}$$

where τ is a reciprocal lattice vector, and the angle $\varepsilon = \frac{\pi}{2} - \theta$ expresses the deviation from perfect backscattering. The parameters τ, ε and k have the uncertainties $\delta\tau$, $\delta\varepsilon$ and δk, respectively. $\delta\tau$ is given by the extension of the reciprocal lattice point parallel to τ and $\delta\varepsilon$ is given by the angular divergence of the beam, which produces an uncertainty of δk and, consequently, an uncertainty in energy as well. The energy resolution given by (2.13) can be written as

$$\frac{\delta E}{E} = \sqrt{\left(\frac{\delta k}{k}\right)_\tau^2 + \left(\frac{\delta k}{k}\right)_\varepsilon^2} \, , \tag{2.15}$$

with the addition of the squared relative uncertainties of k, due to the influences of τ and ε. This approximation allows one to study the different influences.

2.3.1 Crystal Contribution

The contribution $(\delta k/k)_\tau$ reflects the intrinsic quality of the crystal and is given by

$$\left(\frac{\delta k}{k}\right)_\tau = \frac{\delta\tau}{\tau} \tag{2.16}$$

as derivable from (2.14). This contribution is minimized by using perfect crystals. The exact description of scattering from perfect crystals requires dynamical scattering theory (von Laue 1960; James 1963; Pinsker 1978) and leads to

$$\delta\tau = 16 \cdot \pi \cdot r_0 \cdot \frac{|F(Q)|}{\tau \cdot V} \ , \tag{2.17}$$

where $F(Q)$ is the structure factor and V is the volume of the unit cell.

$\delta\tau$, which also is related to the Darwin width, is a characteristic quantity for a given material and a given reflection. $\delta\tau$ is derived for an infinitely thick crystal in symmetric reflection and describes the range of total reflection; however, real crystals have a finite thickness. To take this into account, a parameter Γ can be introduced (James 1963):

$$\Gamma = \frac{\pi}{2d} \cdot \frac{V}{|F(Q)|r_0} = \frac{\tau}{4} \cdot \frac{V}{|F(Q)|r_0} = \frac{4\pi}{\delta\tau} = 2\pi \cdot t_E \ . \tag{2.18}$$

As the thickness of the crystal approaches Γ, a reflectivity of 1 is obtained at the center of $\delta\tau$; however, since the reflectivity curve, better known as the Darwin curve, has strong side bands, Γ can be regarded as a lower limit for the necessary thickness of a perfect crystal. For the (7 7 7) reflection from Si, $\Gamma = 249\mu$m (see Table 2.3). The parameter Γ is correlated with the extinction length t_E as given in (2.18). This extinction length is defined as the penetration depth, where the X-ray intensity is reduced to $1/e$. In analogy to this, the parameter Γ can also be seen as the penetration depth where the radiation is reduced to about $5/1000$ of the incident value.

In a simplified picture, $\delta\tau$ can be regarded as the "reciprocal of the penetration depth of the radiation", though this interpretation is not entirely correct, because $\delta\tau$ is a characteristic quantity which is obtained only for a crystal of infinite thickness.

The intensity of the scattered radiation varies as

$$I \propto k^3 \cdot \delta\beta \cdot \delta\varepsilon \cdot \delta\tau \propto \frac{\delta\beta \cdot \delta\varepsilon \cdot \delta\tau}{\cos^3 \varepsilon_0} \ , \tag{2.19}$$

where $\delta\beta$ stands for the beam divergence perpendicular to $\delta\varepsilon$. For detailed discussions of the dynamical theory for near back reflection diffraction see Kohra and Matsushita (1972); Brümmer, Höche und Nieber (1979) and Graeff and Materlik (1982).

In Fig. 2.5 the energy resolution δE is shown as a function of the X-ray energy for silicon Bragg reflections in backscattering geometry up to energy values of 30 keV. It was calculated after rewriting (2.17) in the form

$$\delta E = \frac{\delta\tau}{\tau} E_i = \frac{16\pi r_0}{V} \cdot |F(Q)| \cdot \frac{E_i}{\tau^2} \ . \tag{2.20}$$

Values of the energy widths for some reflections of the types $(h\ h\ h)$ and $(h\ 0\ 0)$ are marked in Fig. 2.5. The resolution is better for odd reflections than for even reflections, because $F(Q)$ is smaller due to $F(h\ h\ h) = e^{-W} \cdot f(Q) \cdot \sqrt{2}$ for odd h and $F(h\ h\ h) = e^{-W} \cdot f(Q) \cdot 2$ for even h. e^{-W} is the thermal

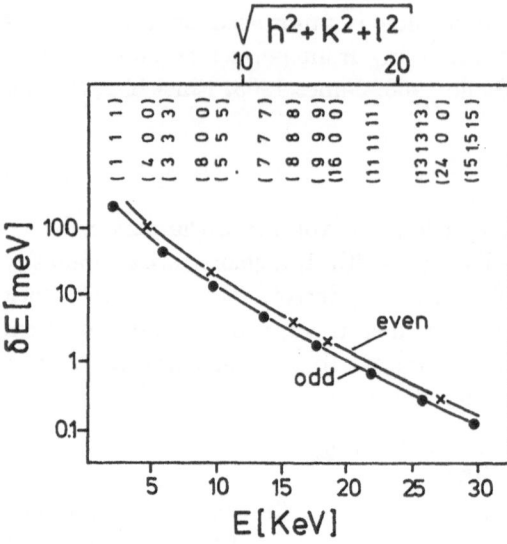

Fig. 2.5. Intrinsic energy resolution of Si as a function of X-ray energy for different $(h\ h\ h)$ and $(h\ 0\ 0)$ reflections in backscattering geometry. The *lines* indicate the variations for reflections with odd ($\bullet$) and even ($\times$) h

Table 2.3. Intrinsic relative and absolute energy resolution of Si for different $(h\ h\ h)$ reflections in backscattering geometry. The values for the photon wavelength λ_i, the energy E_i and the reciprocal lattice vector τ are given as well. The parameter Γ was calculated according to (2.18) and the absorption length $1/\mu$ is taken from Hildebrandt (1979)

Reflection		(3 3 3)	(4 4 4)	(5 5 5)	(7 7 7)	(8 8 8)	(9 9 9)	(11 11 11)
λ_i	[Å]	2.090	1.568	1.254	0.896	0.784	0.697	0.570
E_i	[keV]	5.93	7.91	9.89	13.84	15.82	17.79	21.75
τ	[Å]$^{-1}$	6.01	8.02	10.02	14.03	16.03	18.04	22.04
$\delta\tau/\tau$	[10^{-6}]	9.0	5.1	1.5	0.36	0.28	0.11	0.036
$(\delta E)_\tau$	[meV]	53.4	40.3	14.8	5.0	4.4	2.0	0.8
Γ	[μm]	23	31	84	249	280	633	1584
$1/\mu$	[μm]	29	65	126	351	516	724	1311

Debye-Waller factor and $f(Q)$ is the atomic form factor.

Figure 2.5 clearly demonstrates that an energy resolution better than 10 meV can be achieved by using a Si-reflection of the type $(h\ h\ h)$, for $h \geq 7$. This is also shown by Table 2.3 with selected values of the intrinsic energy resolution of Si in backscattering geometry. However, thus far the influence of the ordinary photo absorption, was completely neglected. If the penetration depth due to absorption is smaller than the extinction length the theoretically predicted energy resolution will not be achieved. Table 2.3 shows the absorption length for Si, taken from Hildebrandt (1979). The comparison with the parameter Γ proves that the absorption influence can be neglected in the discussion of the energy resolution, at least for the discussed reflections.

The photon flux of the source at the needed energy and the quality, i.e. the degree of perfection, of the monochromator and analyzer crystals determine the feasibility of a backscattering experiment with ultra high energy resolution.

2.3.2 Scattering Geometry and Crystal Optics

Basic Equations. According to (2.15) the second important contribution to the uncertainty in the energy resolution is $(\delta k/k)_\varepsilon$, which arises from the variation of the scattering angle and is, therefore, determined by the geometry of the scattering process.

The corresponding wavevector for a beam with scattering angle $\varepsilon = \varepsilon_0 + \delta\varepsilon$ is $k = k_0 + \delta k$, with $k_0 = k(\varepsilon_0)$. This deviation can be derived from (2.14) using a Taylor expansion:

$$\left(\frac{\delta k}{k}\right)_\varepsilon = \tan\varepsilon_0 \cdot \delta\varepsilon + \left(2\tan^2\varepsilon_0 + 1\right) \cdot \tfrac{1}{2}\delta\varepsilon^2 + \dots \tag{2.21}$$

For $\varepsilon_0 \ll 1$ (2.21) can be further expanded. By neglecting higher order terms one obtains:

$$\left(\frac{\delta k}{k}\right)_\varepsilon = \varepsilon_0 \cdot \delta\varepsilon + \tfrac{1}{2}\delta\varepsilon^2 \ . \tag{2.22}$$

The resolution is best when ε_0 and $\delta\varepsilon$ are minimized, which can be achieved by using backscattering geometry.

Equation (2.22) is further transformed and the variation of the wavevector k is written as a function of $\delta\varepsilon$:

$$k = \tfrac{1}{2}k_0 \cdot (\varepsilon_0 + \delta\varepsilon)^2 + k_0 \cdot (1 - \tfrac{1}{2}\varepsilon_0^2) \ . \tag{2.23}$$

For simplifying the further discussion, it is recalled that $\varepsilon = \varepsilon_0 + \delta\varepsilon$. Figure 2.6 shows the variation of the wavevector k as a function of this angle ε for a certain value of k_0.

If $\delta\tilde{\varepsilon}$ is the maximum angle of deviation from ε_0 ($\delta\tilde{\varepsilon} > 0$), then the variation of the angle ε leads to a variation of the scattered wavevector between k_{max} and k_{min} and the influence on the energy resolution is calculated by

$$\frac{\delta E}{E} = \frac{k_{max} - k_{min}}{k} \ . \tag{2.24}$$

The determination of the values k_{max} and k_{min} is illustrated by Fig. 2.6 showing the situations for small deviations ε_0 from backscattering. There is no truly backscattered beam within the scattering range defined by $\delta\tilde{\varepsilon}$ in case Fig. 2.6a in contrast to case Fig. 2.6b where it is part of the scattered beam. As is demonstrated in Fig. 2.6 the determination of k_{min} is different. In case

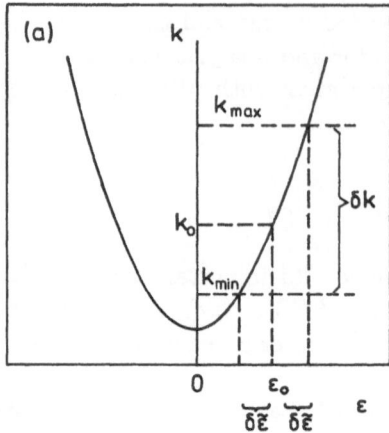 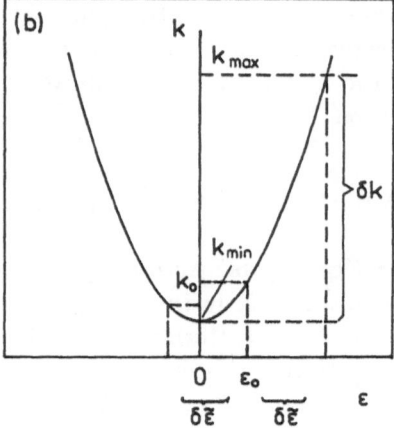

Fig. 2.6a,b. The variation of the wavevector k is shown as a function of the angle ε for a certain value of k_0, according to (2.23), as a measure of the deviation from perfect backscattering. There is no truly backscattered beam within the scattering range defined by $\delta\tilde{\varepsilon}$ in case (**a**), in contrast to case (**b**) where it is part of the allowed range

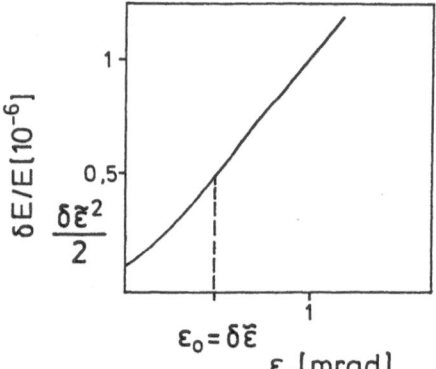

Fig. 2.7. The influence of the scattering geometry on the energy resolution is shown as a function of the deviation of the scattering angle from perfect backscattering. The diagram is given for a typical value of $\delta\tilde{\varepsilon} = 0.5$ mrad

Fig. 2.6b the minimum of the wavevector is determined by the minimum of the parabola. This leads to the following result for the energy resolution :

$$\frac{\delta E}{E} = \begin{cases} 2\varepsilon_0 \cdot \delta\tilde{\varepsilon} & \text{if } \varepsilon_0 \geq \delta\tilde{\varepsilon} \\ \frac{1}{2} \cdot (\varepsilon_0 + \delta\tilde{\varepsilon})^2 & \text{otherwise .} \end{cases} \tag{2.25}$$

Figure 2.7 shows the dependence of the energy resolution on the deviation ε_0 of the scattering angle from perfect backscattering. Smaller angle deviations from backscattering than $\delta\tilde{\varepsilon}$ lead to a parabolic variation of the energy resolution, as long as the direct backscattered beam is within the scattering range. Larger deviations lead to a linear change of the energy resolution. In the case of true backscattering with $\varepsilon_0 = 0$ the resolution is determined by

$$\left(\frac{\delta k}{k}\right)_\varepsilon = \frac{1}{2}\delta\tilde{\varepsilon}^2 . \tag{2.26}$$

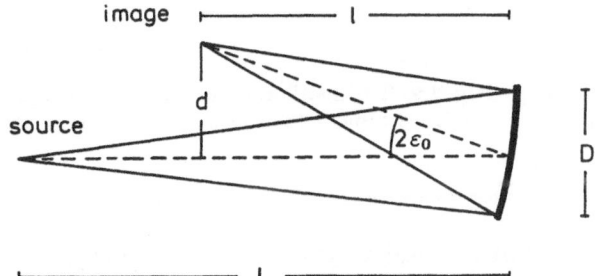

Fig. 2.8. Imaging of a point source by a crystal with demagnification ratio l/L. D is diameter of the crystal and d the distance of the image position from the primary beam

Contribution Due to Demagnification. In order to perform a successful scattering experiment, the experimental arrangement described so far has to be combined with focusing devices for increasing the photon flux. In this case, spherically bent crystals are used to focus the X-ray beam. The position of the image of a point source of X-rays focused by a bent crystal with radius R (Fig. 2.8) is described by

$$\frac{2}{R} = \frac{1}{L} + \frac{1}{l} , \qquad (2.27)$$

with L and l being the distances from the crystal to the source and to the image, respectively. The angle ε_0 is given by $\varepsilon_0 = d/2l$, with d being the distance of the image position from the primary beam.

The source is demagnified by the ratio l/L. For the indicated case of $l < L$ the variation of the angle ε due to the imaging geometry can be described by $\delta\tilde{\varepsilon}_i$, the maximum deviation from ε_0. It can be expressed as

$$\delta\tilde{\varepsilon}_i = \frac{D}{4L} \cdot \left(\frac{L}{l} - 1\right) , \qquad (2.28)$$

where D is the diameter of the spherical crystal.

This deviation leads directly to the corresponding energy resolution, described by (2.25).

One special case should be noted. If l approaches L, i.e. 1 : 1 imaging, then $\delta\tilde{\varepsilon}_i = 0$ and $(\delta E/E)_\varepsilon = 0$, regardless of the crystal dimension D and the angle ε_0 provided that $d \ll L$.

In a scattering experiment with spherical focusing it is not sufficient to discuss the energy resolution within only one scattering plane, such as the plane with ideal backscattering geometry, which is the horizontal one at the instrument INELAX. Additional information from another scattering plane, i.e. the vertical one, is important for describing the complete energy resolution. In this plane there is always deviation from backscattering. This is different from the case of a neutron scattering experiment, where direct backscattering can be achieved in all scattering planes, and can be explained by the much higher velocity of the photons.

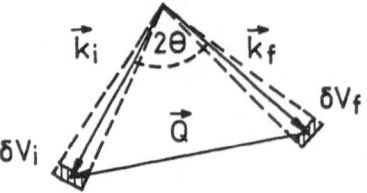

Fig. 2.9. The distribution of photons in reciprocal space around the mean wavevectors k_i and k_f

Contribution Due to Finite Source and Crystal Size. The energy resolution is also affected on the one hand by the finite size h_s of the source and on the other by the finite sizes of the monochromator and analyzer elements. These are blocks with edges of length w obtained by grooving the crystals (see also Sect. 2.5).

These contributions lead to the divergence

$$\delta\tilde{\varepsilon}_c = \frac{\sqrt{h_s^2 + w^2}}{2L} \ .$$

(2.29)

The energy resolution is then given by replacing $\delta\tilde{\varepsilon}$ in (2.25) by $\sqrt{\delta\tilde{\varepsilon}_i^2 + \delta\tilde{\varepsilon}_c^2}$. As already stated, this discussion of the energy resolution gives only a rough estimation. A complete analysis of the geometrical contributions, for instance, has to take into acount the exact value of the scattering angle within the resolution volume and has to determine the specific contributions at each positon on the crystals, which leads to numerical integration.

2.4 Momentum Resolution

For a complete discussion of the resolution of a three axis spectrometer the uncertainty in the momentum transfer $\hbar\delta Q$ has to be taken into account. A detailed discussion of the resolution function in the neutron case is given by Dorner (1972). This function can also be applied to X-rays.

Figure 2.9 shows the distribution of the wavevectors of the photons in an inelastic scattering process within the scattering plane of the reciprocal space. The magnitudes of the areas around the mean wavevectors k_i and k_f depend essentially on the beam divergence, which is controlled by the sizes of the slits and by the illuminated areas on monochromator and analyzer. The folding of both contributions in all three dimensions leads, together with the energy resolution discussed above, to the complete resolution function $R(Q,\omega)$.

It is known from neutron scattering (Cooper and Nathans 1967) that the function $R(Q,\omega)$ has a certain inclination in the (Q,ω) space. But for the case of backscattering instruments (Alefeld 1972; Heidemann 1975), a zero inclination of $R(Q,\omega)$ is achieved by using Bragg angles of 90° at monochromator and analyzer. For X-rays, near backscattering always has a zero inclination of $R(Q,\omega)$.

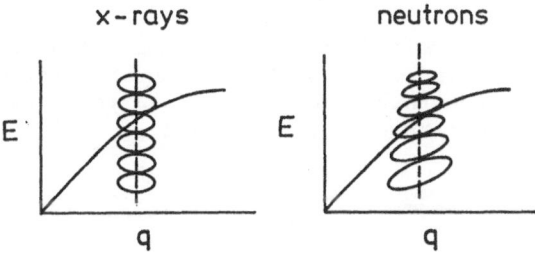

Fig. 2.10. The behavior of the resolution volume in constant-Q scans for the measurement of a phonon dispersion branch with neutrons and X-rays shown in an arbitrary Brillouin zone

Figure 2.10, showing phonon dispersion curves in an arbitrary Brillouin zone, illustrates constant-Q scans for the standard neutron case and for X-rays with phonons to be observed at certain wavevectors q. Besides the different inclination, the size of the resolution volume varies for neutrons and is constant for X-rays. This is due to the strong variation of the wavevector k_f with energy transfer. In the X-ray experiment the temperature variation of the analyzer has no influence on the resolution function.

2.5 Focusing Elements

The measured intensity $I(\boldsymbol{Q}, \omega)$ at a given instrument position corresponding to a momentum transfer $\hbar \boldsymbol{Q}_0$ and an energy transfer $\hbar \omega_0$ is the convolution of the scattering function $S(\boldsymbol{Q}, \omega)$ (2.6) with the resolution function $R(\boldsymbol{Q}, \omega)$:

$$I(\boldsymbol{Q}_0, \omega_0) \propto \int S(\boldsymbol{Q}, \omega) R(\boldsymbol{Q} - \boldsymbol{Q}_0, \omega - \omega_0) \, d^3 Q \, d\omega. \tag{2.30}$$

Even for the high photon flux of insertion devices, it is essential for inelastic X-ray scattering to use focusing optics for both the monochromator and the analyzer crystals; however, bent crystals degrade the energy resolution of the instrument. The influence of the modified scattering geometry on the energy resolution was already discussed in Sect. 2.3.2.

The influence on the intrinsic energy resolution of the crystal is more severe, since perfect crystals are the most important requisite for inelastic scattering. Therefore, it is extremely important to reduce the strains which are induced by bending the crystals. This is done by grooving the crystals in such a way that a thin backwall takes up the strains and leaves volume elements of perfect material on top (Fig. 2.11). The X-rays may not penetrate into the region where backwall strains still have influence. Therefore, the height of the volume elements must be chosen appropriately. In order to get information on the strain distributions in the facets of the spherically bent crystals and to find their best geometrical dimensions, model calculations were made using finite

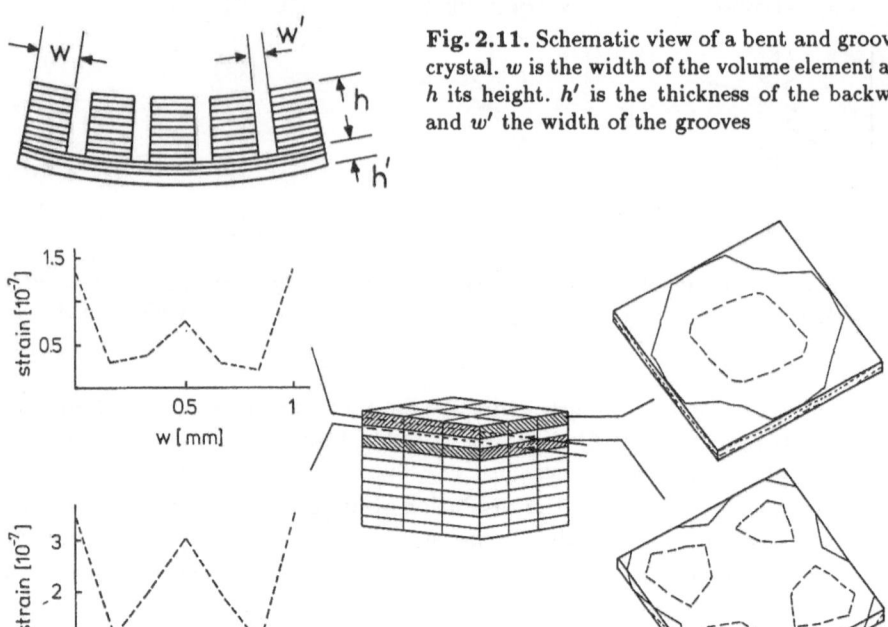

Fig. 2.11. Schematic view of a bent and grooved crystal. w is the width of the volume element and h its height. h' is the thickness of the backwall and w' the width of the grooves

Fig. 2.12a - c. Strain distributions in a volume element of a faceted and spherically bent crystal. (a) shows the strain distributions along lines through the volume element at different heights and (c) illustrates the strain distributions in different slices of the volume element. (b) demonstrates the selected element arrays for the finite element analysis

elements analysis (Strohmeier 1988; Hotz 1991) with the software package "SOLVIA 90"(Solvia, 1987) at the Leibnitz-Rechenzentrum München.

The crystal configuration was simulated by 112 volume elements on top of a thin baseplate. The volume elements were divided into solid elements and the baseplate was divided into 768 shell elements. Rigid-link displacements provided the coupling of the solid elements to the shell elements. The disk was bent by applying a load on the shell elements. The load was varied until a uniform curvature was obtained. The four center volume elements were described by $3 \times 3 \times 10$ separate solid elements with 27 nodes within each element (Fig. 2.12b). The 49 nodes at the base of the center volume element and the ones of the 12 neighboring volume elements were all connected to the baseplate. However, the other volume elements were treated with less resolution.

The displacements of the nodes normal to the baseplate were calculated for the center volume elements. For the determination of the strains along this direction the displacements of the nodes within each element were interpolated before the derivative was calculated. The amplitude of the resulting

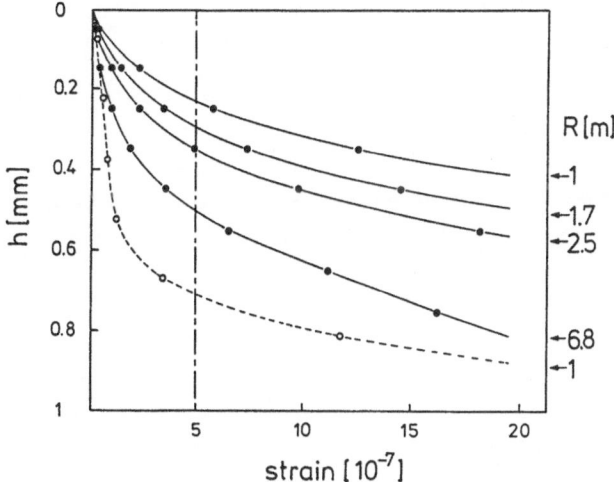

Fig. 2.13. Height dependence of the strains in a volume element of a faceted and spherically bent crystal for different bending radii R, as calculated by finite element analysis for 1 mm (•) and 1.5 mm (○) high volume elements on a 0.1 mm backwall. The value of $5 \cdot 10^{-7}$ for the strain is marked by a *dashed line*. The *lines* are drawn to guide the eye

strains is demonstrated in Fig. 2.12a showing lines through the volume elements at different heights. Figure 2.12c gives the strain distributions with contours of equal strains in thin slices of the volume elements for different heights.

The values of the strains were calculated, in analogy to Fig. 2.12, for the overall height. This allowed a determination of the maximum strain as a function of the height position within the volume element. The result is shown in Fig. 2.13 for the bending radii of 1, 1.7, 2.5 and 6.8 m and volume elements of 1 mm height on top of a 0.1 mm thin disk. It clearly demonstrates the influence of the bending radius on the strain distributions. The thickness of the top layer with strains less than a certain value, for instance $5 \cdot 10^{-7}$ as marked in Fig. 2.13, can be compared with the penetration depth of the X-rays or the crystal thickness Γ, as discussed in Sect. 2.3.1. This rough comparison allows us to test the influence of the bending on the energy resolution. For example, the energy resolution of X-rays of 13.8 keV ($\Gamma = 0.25$ mm, Sect. 2.3.1) will not be influenced by a bending radius of 2.5 m. X-rays with higher energies will penetrate deeper into the crystal. In this case, a more sophisticated analysis has to be performed to see the percentage of the crystal volume having strains with values more than a certain critical value.

The influence of an increased cube height of 1.5 mm on top of the volume element of the same thin (0.1 mm) backwall is also shown in Fig. 2.13 for a bending radius of 1 m. It demonstrates the gain of additional height for the scattering process in the top part of the volume element.

3. X-ray Sources

As already mentioned, the development of the new sources for X-rays of high intensity has stimulated attempts to perform inelastic scattering experiments with high energy resolution. In a brief summary the main features of the conventional and the new X-ray sources are discussed in the following sections. For detailed information, particularly on the new sources, the reader is referred to Koch, Eastman and Farge (1983) and Gudat (1987) and for the basic theory to Schwinger (1949) and Jackson (1962).

3.1 Conventional X-ray Generator

X-rays are created in two different ways. High energy electrons slowed down in an anode material give rise to radiation with a continuous frequency spectrum. This radiation is called bremsstrahlung. It is observed when any charged particle is either accelerated or decelerated. The total radiated power P of an accelerated electron was calculated by Larmor to be

$$P = \frac{2e^2}{3c^3} \cdot \left| \frac{d\boldsymbol{v}}{dt} \right|^2 , \tag{3.1}$$

where $\boldsymbol{v}$ describes the velocity of the electron. The angular distribution of the radiated power is given by

$$\frac{dP}{d\Omega} = \frac{e^2}{4\pi c^3} \left(\frac{d\boldsymbol{v}}{dt} \right)^2 \cdot \sin^2 \varphi \tag{3.2}$$

with φ being the angle between the vectors of acceleration and radiation. This distribution has rotational symmetry.

The second process which can be used for X-ray generation is the recombination of electrons after creation of unoccupied states in the inner shells of atoms in the anode by collisional and/or photoionisation. This recombination radiation is emitted as a sharp characteristic line with a wavelength determined by the transition energy between the corresponding shell levels.

The effectiveness of an X-ray tube is very low because less than 1% of the absorbed electron energy is emitted as X-ray radiation and the dominant part is converted to heat. Therefore, the X-ray intensity of conventional tubes is limited by the technical possibilities for the cooling of the anode. X-ray generators with water-cooled and rotating anodes allow working with electric powers of up to 90 kW.

3.2 Synchrotron Radiation

Synchrotron radiation is the radiation emitted by charged particles which are accelerated on circular orbits in synchrotrons or storage rings. Due to the high energy E_{el} of the particles, in the range of GeV, and the corresponding high velocity, close to the velocity of light, ($v/c \approx 1$) the relativistic version of (3.1) has to be used to calculate the emitted radiation:

$$P \simeq \frac{2e^2c}{R^2} \cdot \left(\frac{v}{c}\right)^4 \cdot \left(\frac{E_{el}}{mc^2}\right)^4 \sim \frac{\gamma^4}{R^2} , \tag{3.3}$$

where R is the radius of the storage ring or synchrotron and m the rest mass of the particle. Due to the factor $(mc^2)^{-4}$ only electrons and positrons lead to strong radiation. The power of the emitted radiation is proportional to the fourth power of the reduced energy ($\gamma = E_{el}/mc^2$). Therefore, with $mc^2 = 511$ keV for electrons, γ is of the order of 10^4.

The angular distribution of radiation for an accelerated charge in extreme relativistic motion is no longer described by the dipolar distribution of (3.2) and Fig. 3.1a. The relativistic transformation causes a distortion of the radiation pattern in the direction of motion, as shown in Fig. 3.1b. It becomes a narrow cone in the direction of the instantaneous velocity vector of the charge. Radiation will be visible only when the velocity of the electron is directed to the observer. It will be a burst of radiation, very short in time with $\Delta t \sim (R/c) \cdot \gamma^3$. The Fourier transform of this pulse shows a corresponding broad frequency spectrum of white radiation with a spectral cutoff just above the critical energy E_c, used to characterize the spectrum (see also Fig. 3.5). It is defined in such a way that half of the total power is radiated above the critical energy and half below. The critical wavelength λ_c is defined, correspondingly.

The angular width of the beam is given by

$$\delta\psi \simeq \frac{1}{\gamma}\left(\frac{\lambda}{\lambda_c}\right)^{1/2} \text{ for } \lambda \ll \lambda_c \text{ and by } \delta\psi \simeq \frac{1}{\gamma}\left(\frac{\lambda_c}{\lambda}\right)^{1/3} \text{ for } \lambda \gg \lambda_c. \tag{3.4}$$

(3.4) demonstrates that the angular width is strongly dependent on the energy of the radiation.

The horizontally moving electrons in a circular orbit can emit radiation at any point of the orbit within the bending magnets. Thus, the natural angular width of the radiation cone can only be observed in the vertical direction, leading to a vertical divergence of the photon beam. The horizontal angular width depends on the length of the observed arc (see also Fig. 3.4a).

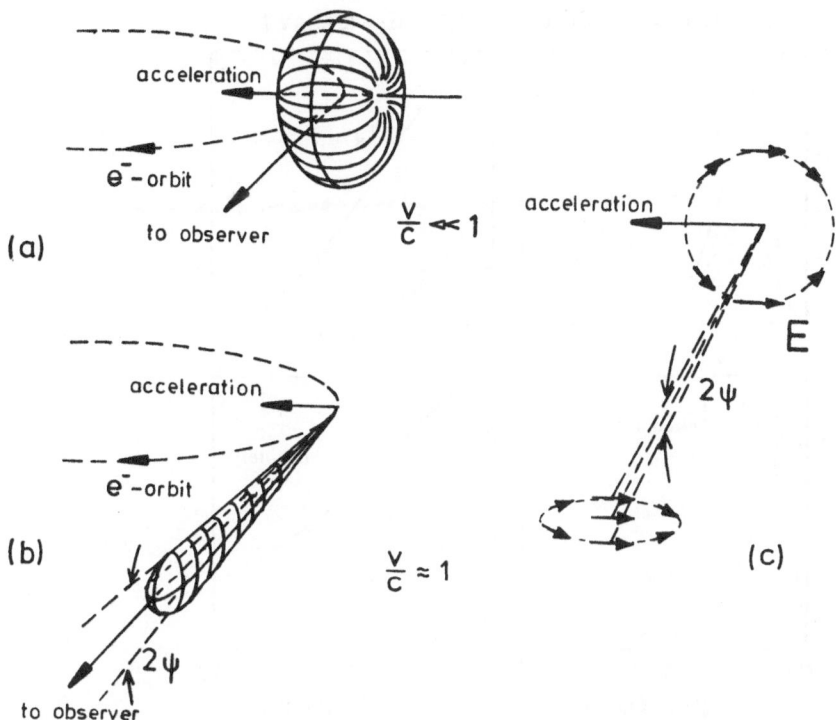

Fig. 3.1. Radiation distribution for an electron for nonrelativistic (a) and relativistic (b) movement on a circular orbit. (c) shows the far field E distribution with the dipole-characteristic in a plane vertical to the orbit plane. The distributions in the rest frame and after Lorentz transformation in the lab frame illustrate the polarization of the synchrotron radiation

There is another important consequence of the relativistic motion of the electron or positron. The electric field E with its characteristic dipole distribution in the rest frame is contracted by the Lorentz transformation (Fig. 3.1c). As a consequence, the radiation is dominantly polarized with E in the plane of motion. As long as the radiation is observed within the orbit plane, it is completely linearly polarized. Observation at a small elevation angle ψ out of this plane will detect elliptically polarized radiation.

Figure 3.2 demonstrates the angular dependence of the polarization of the radiation for two photon energies E_i with DORIS working at 3.5 GeV. It shows the percentage of the photon flux for the polarization components parallel and vertical to the orbit plane, as well as the percentage of the flux being linearly and circularly polarized. This diagram also illustrates the influence of the radiation energy on the vertical divergence of the beam as discussed in connection with (3.4). The spectrum of radiation depends strongly on the energy of the moving electron or positron. In storage rings, compared to synchrotrons, the particles are kept at a certain energy. The energy loss

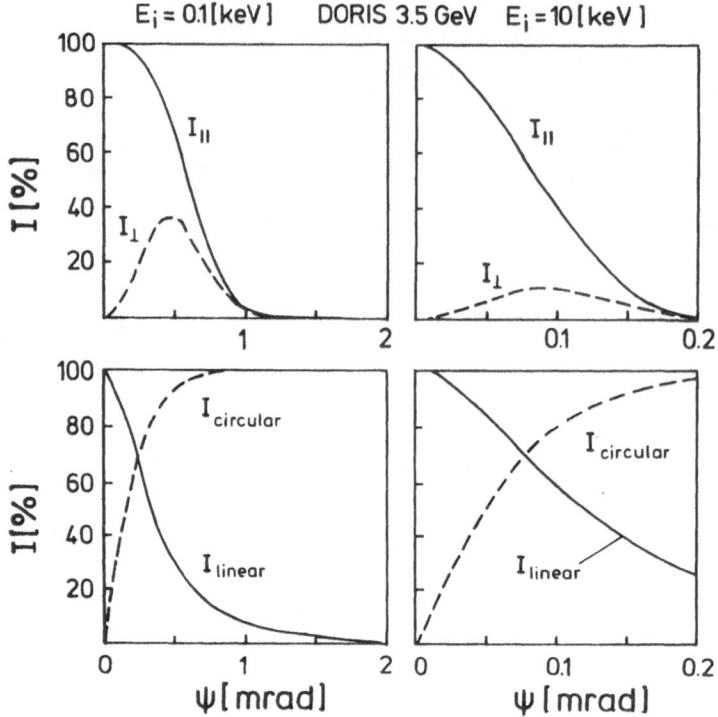

Fig. 3.2. Variation of the polarization components parallel (∥) and vertical (⊥) to the orbit plane with the elevation angle ψ out of the plane, as well as the variations of linearly and circularly polarized components. The photon flux is given as the percentage of the flux at $\psi = 0$

of the particle due to the emitted radiation is compensated by steady high frequency energy input. Only particles having the correct orbit frequency will be accelerated in the right phase. This leads to the typical bunch structure of the electron or positron beam. The bunch length for the storage ring DORIS II at DESY, for example, is about 150 ps; single and multiple bunch modes are possible.

For a detailed discussion of the exact spectral distribution of the synchrotron radiation it is referred to the literature, as already mentioned. Here, only the frequently used quantities for the description of the photon flux, important for applications, will be given. The differential photon flux is normally called spectral brightness and has the dimension [photons/(s · mrad^2·0.1% $\delta E/E$)]. The average spectral brilliance of the photon beam takes additionally the source size into account and its dimension is [photons/(s · mm^2· mrad^2 · 0.1% $\delta E/E$)]. Both quantities can be given per mA or partially integrated over a given storage ring current.

The possibility to increase the photon flux of a given storage ring by the installation of insertion devices will be discussed further.

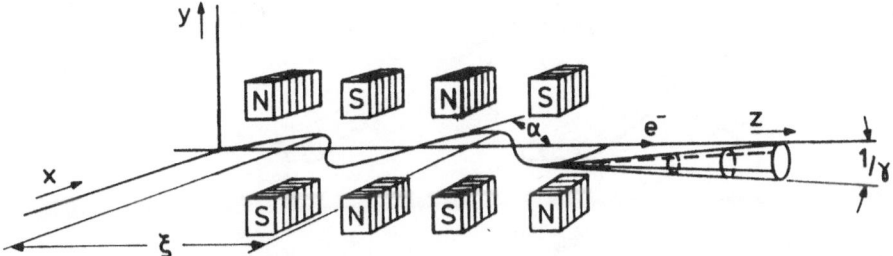

Fig. 3.3. Transverse multipole magnetic structure with radiation emission for relativistic electrons. α is the maximum deflection angle, $1/\gamma$ the natural opening angle of the radiation cone and ζ the period length of the magnetic structure

3.3 Insertion Devices

Insertion devices are periodic magnetic structures that are installed in straight sections of storage rings to get additional periodic deflections of the electron beam without disturbing the rest of the orbit. The additional oscillatory movement causes radiation. Its spectral and spatial behavior is determined by the specific magnetic structure. Figure 3.3 demonstrates the radiation emission for relativistic electrons in a transverse multipole magnetic structure with a certain period length ζ. The maximum deflection angle α and the natural opening angle $(1/\gamma)$ of the radiation cone are indicated.

Figures 3.4b, c, d indicate the electron beam trajectories and the radiation patterns for the insertion devices wavelength shifter, wiggler and undulator. For comparison, a bending magnet is given as well (Fig. 3.4a). In all cases the vertical angular width of the emitted radiation is of the order of $1/\gamma$. Only the horizontal angular width is varying.

A wavelength shifter (Fig. 3.4b) essentially locally changes the radius of the electron path of a given storage ring. Because the power of radiation and its frequency distribution depend on the radius, their characteristics can be varied, particularly the critical energy.

An undulator (Fig. 3.4c) is a spatial periodic magnetic structure which leads to the emission of quasi-monochromatic radiation of relativistic electrons with a strongly increased photon flux. The magnetic structure leading to such strong interference effects has to produce beam deflections which are always smaller than the opening angle $(1/\gamma)$ of the radiation. Therefore, the horizontal angular width of the radiation is about $1/\gamma$ for $\lambda \simeq \lambda_c$. The brightness of the radiation emitted by an undulator is greater than for an ordinary bending magnet. This ratio is about the square of the number of periods. Due to the source size of small emittance rings the brilliance of the undulator is very high.

Periodic magnetic structures leading to stronger deflections of the trajectory are called wigglers (Fig. 3.4d). Such multipole wigglers produce a complicated spectrum, which on the high energy side resembles the simple addition

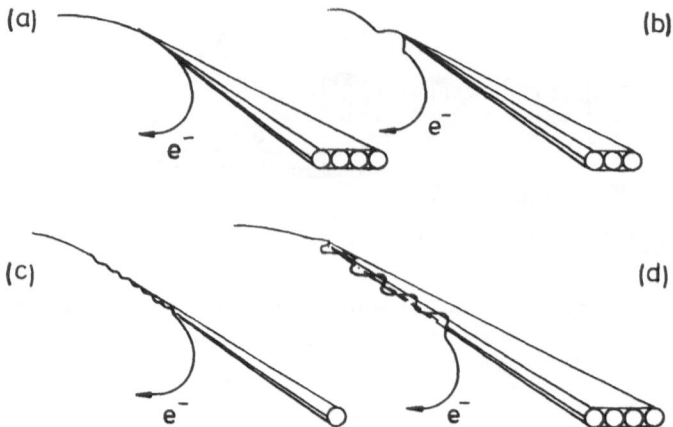

Fig. 3.4. Beam trajectories and emission patterns for bending magnet (a), wavelength shifter (b), undulator (c) and wiggler (d).

of the spectra of several bending magnets. The brightness is increased by a factor corresponding to twice the number of periods. The horizontal opening angle and, therefore, the beam divergence is larger for the wiggler than for the undulator.

Special magnetic structures can also be defined to generate circularly polarized radiation. One possibility is to install a crossed undulator (Moissev, Nikitin and Fedorov 1978) to obtain the desired polarization by the interference between two crossed sections. Another way is the use of an asymmetric wiggler (Goulon, Ellaume and Raoux 1987). In this structure the magnets are arranged in such a way that the positive and negative half-waves of the field profile have different amplitudes and are no longer symmetric. Nevertheless, a straight trajectory is maintained because the integral within a single period is still zero. The emitted radiation can be described by two different critical energies determined by the field profile. In the photon energy range between these two critical energies circularly polarized radiation will be observed off axis.

In concluding this brief chapter on X-ray sources, Fig. 3.5 illustrates photon intensities of online devices. The brightness of a bending magnet of the DORIS II storage ring at DESY, Hamburg, and the photon brightness of the hard X-ray wiggler HARWI at DORIS II are compared to the brightness produced by an X-ray generator with a rotating anode (90 kW) for the characteristic K_α line of the anode materials Ag, Mo, Cu and Fe and the Bremsstrahlung of Mo. The values are either taken from Bonse 1979 or calculated with the equations given there. In the case of the characteristic lines the natural linewidth was taken as $\delta E/E$. The spectra of DORIS are calculated for a current of 100 mA at 4.5 GeV operation (HASYLAB 1987).

For the evaluation of the total photon flux for the compared sources, the large useful divergence and the small source size of conventional sources will

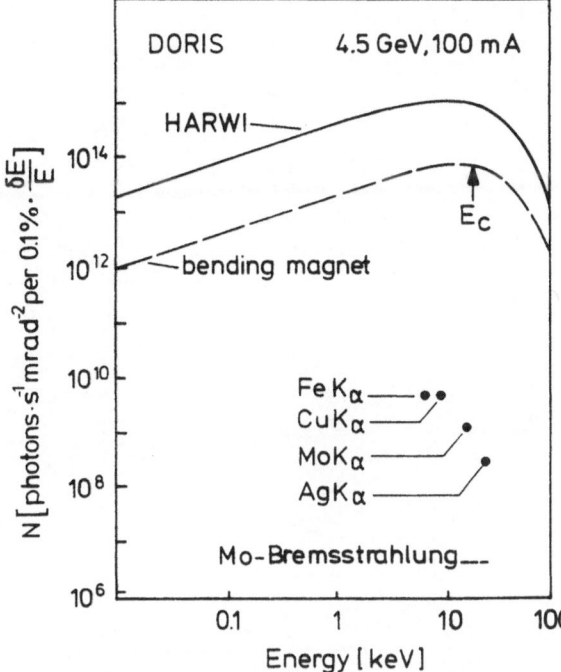

Fig. 3.5. The brightness of a bending magnet and the brightness of the X-ray wiggler HARWI at DORIS II are compared with the brightnesss of an X-ray generator with a rotating anode for the characteristic K_α line of different anode materials and the bremsstrahlung of Mo

in certain applications diminish the differences shown in Fig. 3.5. In any case, the comparison shown in Fig. 3.5 clearly demostrates the progress achieved with the development of synchrotron radiation and insertion devices. Some additional aspects of future developments will be discussed later in Sect. 7.2.

4. The INELAX Instrument

The inelastic X-ray scattering spectrometer INELAX has been built in several stages since 1984 at the DORIS II storage ring of HASYLAB at DESY, Hamburg. Initially, the spectrometer had been installed at a DORIS beam port using the radiation from a bending magnet. It was later moved to the wiggler line W2 (HARWI). The latest set-up is shown schematically in Fig. 4.1.

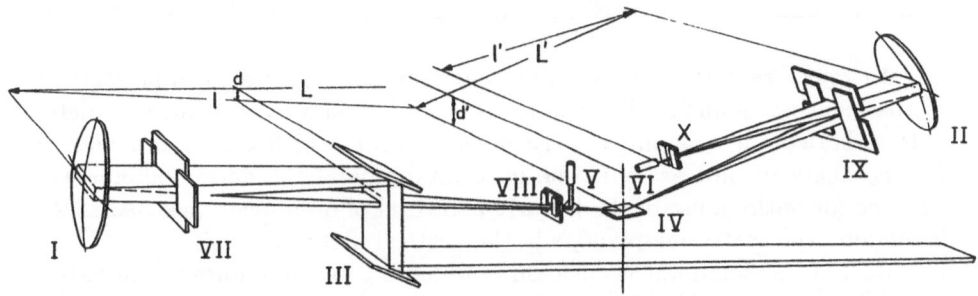

Fig. 4.1. The instrument INELAX at the HARWI wiggler line of HASYLAB. The beam passes the premonochromator (III) and the monochromator (I) before it illuminates the sample (IV). The scattered intensity is focused by the analyzer (II) into the detector (VI). The primary intensity is controlled by a monitor unit (V). (VII-X) indicate slit systems. For lengths of the labeled distances see Table 4.1

The primary parts of the instrument are the monochromator (I) and the analyzer (II) units. Both components consist of spherically bent crystal silicon disks and operate under extreme Bragg backreflection with $\theta = 89.98°$ at the monochromator and $\theta = 89.95°$ at the analyzer. These scattering angles correspond to angular deviations $\varepsilon = 0.375$ mrad and $\varepsilon = 0.80$ mrad from backscattering at the monochromator and at the analyzer, respectively. The photon beam from the storage ring passes the premonochromator (III) on its way to the monochromator. This premonochromator acts essentially as a heat filter to protect the monochromator, which defines the desired small energy bandwidth of the photon beam focused onto the sample (IV). The beam intensity is registered by a monitor device (V), using the scattered intensity from a kapton foil within the beam path.

The analyzer crystal collects the radiation scattered by the sample from a variable solid angle and focuses it onto the detector (VI), which is an ion-implanted Si-diode working just below room temperature. Various slit systems (VII – X) help to define the beam cross section and the scattering geometry

Table 4.1. Geometric distances of INELAX (Fig. 4.1) at the HARWI wiggler at DORIS. The source for the beam to the analyzer is slit VIII

Distances				Monochromator	Analyzer
Source to crystal	L		(m)	38	2.6
Crystal to image	l		(m)	8	2.5
Bending radius	R		(m)	13	2.55
Distance between both beams	d		(mm)	6	4
Beam size at crystal	D	hor	(mm)	90	80
	D	ver	(mm)	8	80
Crystal element size	w	hor	(mm)	1	1
	w	ver	(mm)	1	1
Source size	h	hor	(mm)	3.5	2
	h	ver	(mm)	0.5	1

and to reject radiation at unwanted energies, arising due to imperfections of the focusing elements. The sample is mounted on a special diffractometer with a horizontal axis and a large detector arm, which serves as a mount for the analyzer unit as well. The horizontal axis of the diffractometer was selected for optimal use of the linearly polarized photon beam (see Sect. 3.2). Therefore, the scattering plane is in the vertical plane.

Most of the beam line is enclosed in vacuum tubes to reduce the intensity losses due to absorption along a distance of 51 meters. The relevant geometrical distances of the instrument INELAX are given in Table 4.1. Figure 4.2 gives an impression of the experimental hall of the HARWI wiggler line of HASYLAB. The lead hutches, seen downstream, contain the main diffractometer of the instrument INELAX in the first part and the monochromator in the last part. The control station for the instrument is visible in the foreground.

A view inside the first lead hutch in Fig. 4.3 shows the main diffractometer of the instrument INELAX with the special arm for the analyzer unit. The beam path of the white beam, shielded with lead and covered with aluminum is visible along the hutch downstream of the beam, where it ends in the just-visible lead-shielded premonochromator device.

4.1 The Main Components

4.1.1 Premonochromator

The use of a premonochromator is essential to take the heat load of the white synchrotron beam off the monochromator. Uncontrolled heating of the monochromator would lead to gradients in the lattice parameter, thus destroying the intrinsic energy resolution of the monochromator crystal. By the use of a

Fig. 4.2. A view of the experimental hall of the HARWI wiggler line of HASYLAB. The lead hutches, seen downstream, contain the main diffractometer of the instrument INELAX in the first part and the monochromator in the last part. The control station for the instrument is visible in the foreground

Fig. 4.3. A view inside the first lead hutch showing the main diffractometer of the INELAX instrument with the special arm for the analyzer unit. The beam path of the white beam, shielded with lead and covered with aluminum is visible along the hutch downstream, where it ends in the just-visible lead-shielded premonochromator device

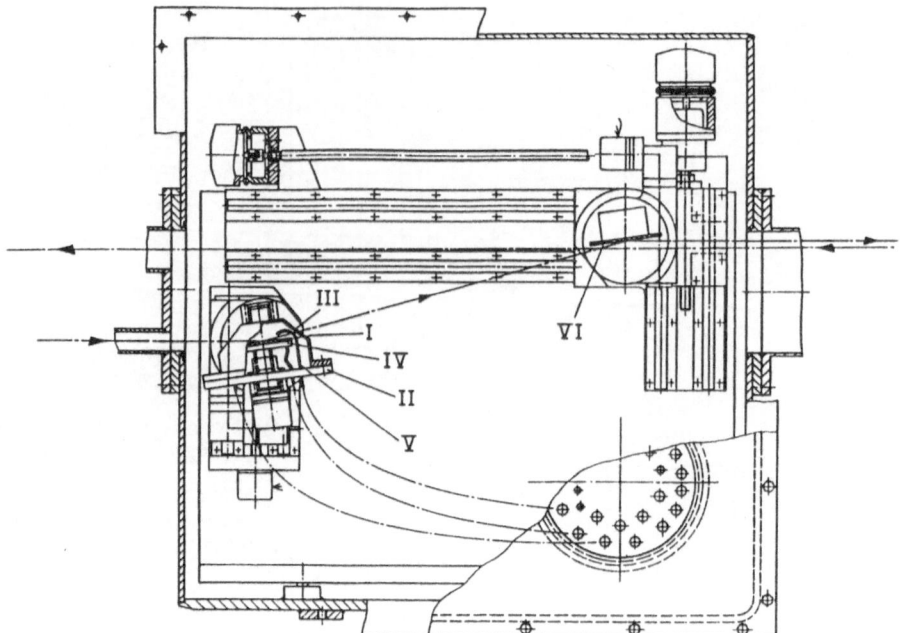

Fig. 4.4. Schematic of the premonochromator arrangement. Both crystals are mounted on turntables and slides for alignment and tuning of the energy E_i. The beam path is indicated as well

premonochromator, the temperature of the monochromator can be well stabilized.

The premonochromator operating in the vertical plane should transmit the full vertical divergence for that wavelength which is diffracted with high resolution by the monochromator. The vertical divergence of a synchrotron beam has two components. One component is $\delta\psi$ as defined by (3.4) and depends on the energy ($\gamma = E_{el}/mc^2$) and on the characteristic wavelength λ_c of the emitting device. The other component arises from the vertical divergence $\delta\psi_{el}$ of the electron beam at the emitting device. For $\lambda \simeq 0.89$ Å, which corresponds to the (7 7 7) reflection of Si (Fig. 2.5), and DORIS operating at 3.7 GeV one obtains a total vertical divergence

$$\delta\psi_{tot} = \sqrt{\delta\psi_R^2 + \delta\psi_{el}^2} \simeq 2 \cdot 10^{-4} \text{ rad} . \tag{4.1}$$

At present, specially tempered silicon crystals with Darwin widths broadened to twice the natural intrinsic width are used in symmetric reflection (1 1 1) as the premonochromator crystals.

The first crystal (I) of the double monochromator (Fig. 4.4) is enclosed in a special housing (II) within the vacuum containment of the complete premonochromator mechanism. The beryllium windows (III) which are installed for the X-ray beam are watercooled. The crystal is mounted on a watercooled copper plate (IV) and is thermally coupled to it by a liquid gallium film. In addition, the top surface of this crystal, which is illuminated by the white wiggler beam, is flooded by streams of gaseous helium (V) to take off the heat load directly at the crystal face and to reduce thermal strains in the top layers of the silicon crystal. This arrangement allowed an intensity gain of a factor of three, because of a better matching of the Darwin curves of both crystals. This gain is similar to the gain obtained by earlier tests of gas cooling (Burkel, Dorner, Illini and Peisl 1990).

Even more intensity gain is expected by installing a silicon crystal, which is additionally cooled by a liquid gallium flow through channels in the crystal, as studied by Smither et al (1989) and Bilderback, Lairson, Barbee, Ice and Sparks (1989). Another promising approach is the cooling of the silicon crystals down to a temperature of 120 K, since the thermal expansion coefficient is about zero around this temperature (Okada and Tokumaru 1984). This would reduce distortions which are normally caused by temperature gradients due to the heat load. At the same time, the thermal conductivity of silicon increases by one order of magnitude when cooling to this low temperature (Bilderback 1986).

The geometrical position of the premonochromator between the sample and the monochromator was selected so that the backscattered beam from the main monochromator passes between the two crystals of the premonochromator on its way to the sample (Fig. 4.1).

This configuration has some major advantages:

– The deviation of the scattering angle from direct backscattering at the monochromator crystal is reduced, thus allowing for better energy resolution, according to (2.25).

– The sample position is lifted out of the plane of the white beam of the storage ring. Therefore, the sample and the detector arm are not positioned in the direct scattering cone of the white beam, which is produced when it hits the first crystal of the premonochromator. Shielding problems can be solved in this configuration more easily and, therefore, the background scattering is reduced.

– The space at the sample position is increased, thus making different sample environments possible.

– In the future, it will be possible to study backscattering with even smaller deviations from the direct backscattering case. Switching from Bragg to Laue diffraction at the second crystal (VI) in (Fig. 4.4) of the premonochromator would transmit the backscattered beam through this crystal. Such Laue crystals are under development at the angiography project (Dix, Glüer, Graeff, Höhne and Kupper 1982) and were discussed for instrumental improvements (Gehrke 1987) at HASYLAB.

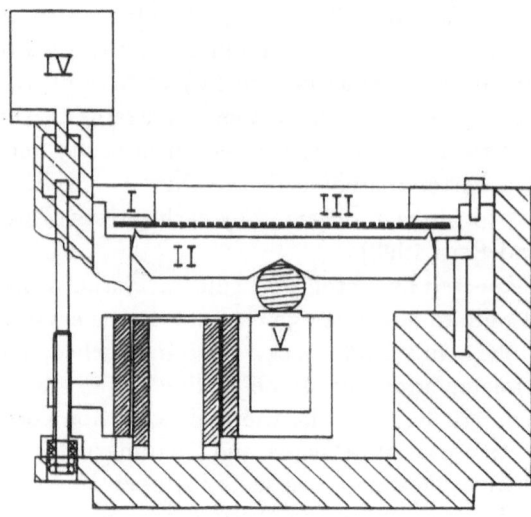

Fig. 4.5. Monochromator housing for alignments to the spherical curvature of the crystal. The crystal (III) is bent by two concentric rings (I) and (II), which can be aligned by a stepping motor (IV) and a piezo crystal device (V)

- This special geometry might further open the interesting area of liquids, since it allows an angle in the vertical plane between the incoming beam and the horizontal surface of the liquid. In this way, reflection measurements of liquids become possible; however, a worse energy resolution must be expected. By using different lattice spacings in the two monochromator crystals this influence can be diminished.

The complete premonochromator arrangement is totally enclosed by lead bricks, so that inelastic measurements can be performed during high energy shifts of the storage ring. The background at the detector position is suppressed to the level of about one count per minute.

4.1.2 Monochromator

The monochromator consists of a crosswise grooved Si-crystal disk. ([1 1 1] normal to the surface). In order to focus the beam onto the sample, this crystal is spherically bent. Because of the variable distance of the monochromator to the sample, it was necesssary to develop a device for spherical bending (Burkel, Dorner, Illini and Peisl 1990) in analogy to Fuji, Hastings, Ulc and Moncton (1982) and Egger, Hofmann and Kalus (1984). It is shown in Fig. 4.5. The Si disk (100 mm diameter) is 1.2 mm thick and grooved to obtain facets of an area of 1×1 mm^2 on a 0.15 mm thick backwall. With the aid of two concentric rings (I, II) the Si disc (III) is bent to the desired radius (Table 4.1). A motor (IV) and a tunable piezo-crystal device is used for rough and fine alignment, respectively.

In addition, another technique has been used lately to obtain a very good focusing monochromator. A grooved silicon disc was attached to a concave glass ring only by adhesion of a thin glycerine film. This arrangement has the

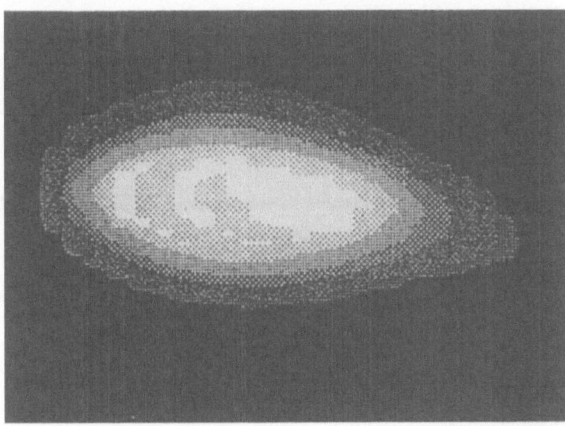

Fig. 4.6. Intensity distribution at the focus of the spherically bent monochromator observed at the wiggler station W2. The dimensions at half intensity are about 3 mm horizontal and 1 mm vertical

advantage that the strained silicon in the grooves of the crystal can be shaded by filling the grooves with an absorbing mixture of glycerine and Ta-powder (Burkel, Dorner, Illini and Peisl 1989).

The performance of the monochromator is demonstrated in Fig. 4.6, which shows a typical intensity distribution obtained by observing the light of a fluorescence screen with a CCD camera. The dimensions at half intensity are about 3 mm horizontal and 1 mm vertical.

4.1.3 Analyzer

At the analyzer position it is not possible to use a bending device because of the temperature variation during the measurements. Therefore, a permanently spherically bent crystal is used (Fig. 4.7). A crosswise grooved Si disk (100 mm diameter) is pressed into a concave glassform and stabilized by glue, a technique which is still under improvement. In the experiments described here, the analyzer had a focus distance of 2.5 m.

Figure 4.8 gives a view of the housing of the analyzer crystal, which allows for stabilization of the temperature to better than ± 0.01 K. Rough control of the crystal temperature is obtained by a closed water cooling circuit (I), which is coupled by a Peltier element (II) to a copper plate (III), containing heating wires for the fine control. The crystal on the glass form (IV) is enclosed in a gas atmosphere for heat exchamge. The angle alignment is done by tilting segments (V). The whole arrangement is in vacuum. The use of the Peltier element has the advantage of additional heating or cooling, according to the directions of the current, if necessary.

Fig. 4.7. Grooved silicon crystal glued to a concave glass form as it is used in the analyzer unit

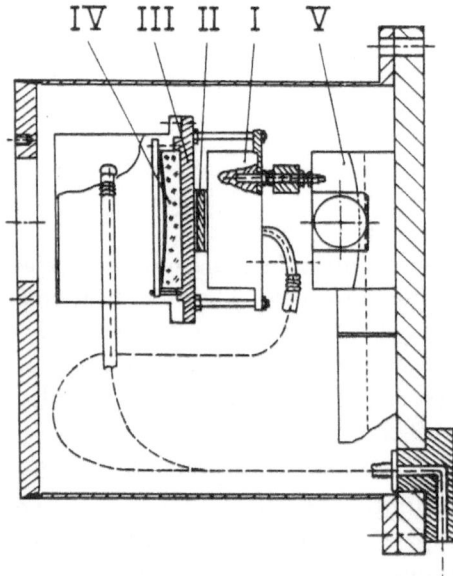

Fig. 4.8. Analyzer crystal housing with temperature stabilizing arrangements.

4.1.4 Overview of the Network of INELAX Instrumentation

There is a wide range of coordination problems that has to be solved for the perfect performance of the instrument and the optimal study of the inelastic scattering processes. Figure 4.9 demonstrates how the different challenges are met. They are briefly summarized in the following.

The long path of the beam and its scattering from four perfect crystals and from the sample require optimal alignment. This includes the stabilization of the premonochromator in energy and the interaction of premonochromator and monochromator. For the control of this interaction an additional small diffractometer unit is installed in the beam path between premonochromator and monochromator. In general, the alignment problems are solved with a variety of mechanical adjustments using stepping motors and piezo-crystal drives and with the help of television cameras for survey jobs.

The stabilizations of the temperatures of the monochromator and analyzer crystals and the controlled variations of their temperatures are most important for the performance of the instrument. As already described, this task is separated into rough and fine alignment of the temperature by water cooling circuits and electric heating devices. The latter are controlled by highly sensitive resistor bridges. Independent control is achieved by observing the frequency changes of quartz sensors with temperature.

Problems like photon counting with single channel or multichannel analysis and general data handling are standard tasks. The complete control of the instrument is done with a computer via a standard data bus.

4.2 The Technique for Inelastic X-ray Measurements with INELAX

As stated in Sect. 2.2 energy scans at the INELAX instrument are performed as constant-Q scans. In these scans all scattering angles are kept fixed and only the temperature of the analyzer (T_{ana}) with respect to the temperature of the monochromator (T_{mono}) is varied. In order to obtain correct values for the energy transfers in the inelastic scattering process, a calibration is necessary.

4.2.1 Energy Calibration

Energy scans at INELAX are done in a dynamic way. This means that the temperature and the energy transfer are changed at a certain speed, defined by the size of the temperature step and by the counting time. Hereby temperature gradients can occur between the temperature of the probe and the analyzer crystal, but using identical scan conditions for the inelastic scans delivers reproduceable and reliable results. The time for temperature stabilization would increase the scan length enormously. Only for larger energy

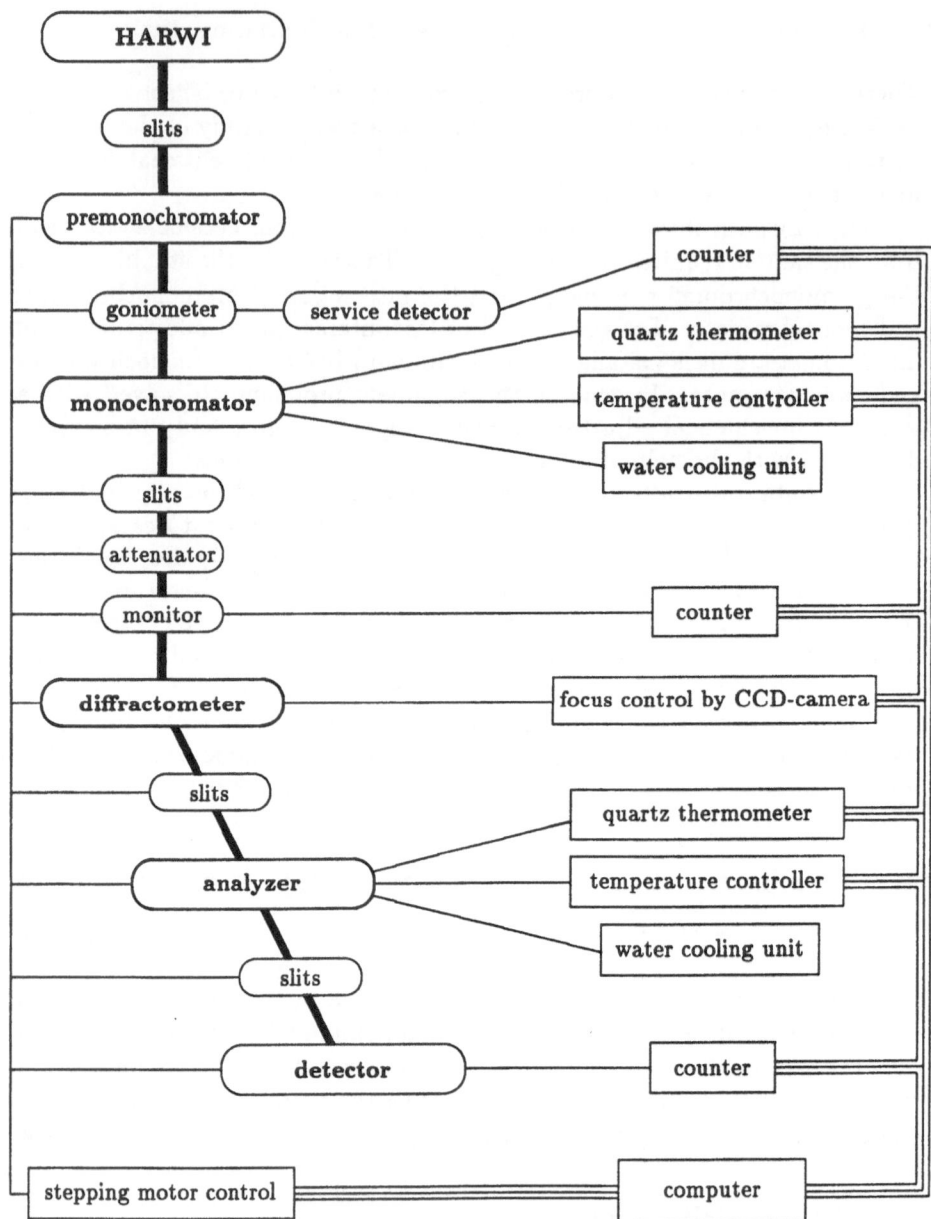

Fig. 4.9. Instrumentation and control units of INELAX

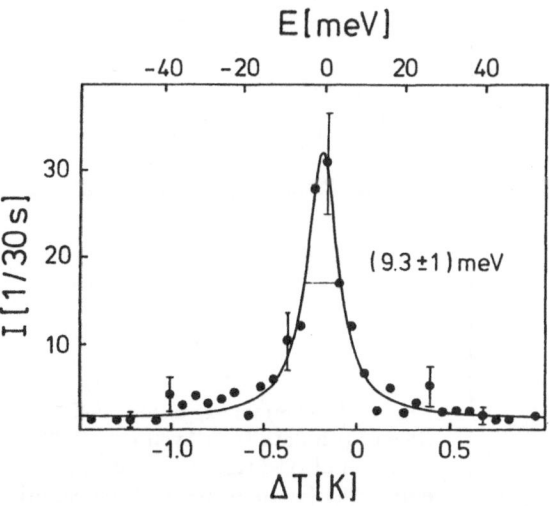

Fig. 4.10. Intensity elastically scattered from fused silica as a function of the temperature difference between the analyzer and the monochromator and the corresponding energy transfer

transfers or if extremely long counting times are necessary, constant-E scans are performed by varying Q (see also Sect. 6.4).

The calibration of the zero energy transfer is made by a constant-Q scan by recording the elastic scattering intensity from an amorphous sample such as fused silica. This scan is always performed with the same speed of temperature variation as the one used for the inelastic scans later. Such a scan of the elastically scattered intensity from fused silica is given in Fig. 4.10 as a function of the temperature difference between the analyzer and the monochromator crystals.

The zero points for energy transfer and the temperature difference between monochromator and analyzer may not coincide. This is caused by the different Bragg angles at the analyzer and the monochromator and, of course, by differences in the temperature sensors. These quantities can be taken into account by a calibration constant c_0. The energy transfer E is correlated with the temperature variation $\Delta T = T_{\text{ana}} - T_{\text{mono}}$ through the thermal expansion coefficient $\alpha(T)$ of Si and through the incident X-ray energy E_i:

$$E = \alpha(T) \cdot E_i \cdot (\Delta T + c_0) \ . \tag{4.2}$$

The maximum of the elastically scattered intensity from the fused silica in Fig. 4.10 corresponds to zero energy transfer and, therefore, its temperature offset determines the calibration factor c_0 to be 0.2 K in this example.

4.2.2 Energy Transfer

After having derived the calibration factor c_0 according to (4.2), the same equation can be used for the calculation of the actual energy transfer E, determined by the temperature difference between monochromator and analyzer. The energy scale shown in Fig. 4.10 is found in this way by using E_i

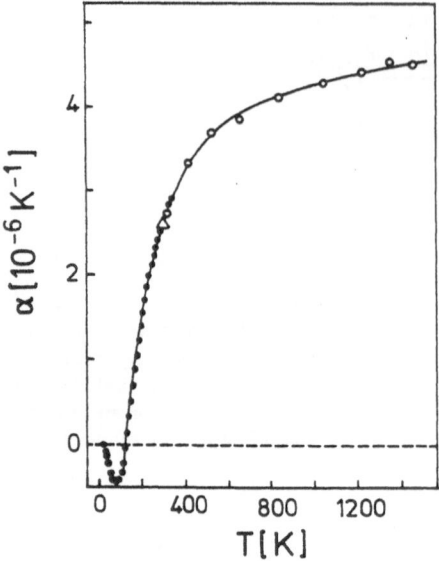

Fig. 4.11. Thermal expansion coefficient of silicon with data from Lyon et al 1977 (•), Becker et al 1982 (△) and Okada and Tokumura 1984 (o) compared with an empirical curve as discussed in the text

= 17.8 keV (Fig. 2.5) and the thermal expansion coefficient for high purity silicon, which is well known for room temperature (Becker, Seyfried and Siegert 1982), $\alpha(300\,\mathrm{K}) = 2.6 \cdot 10^{-6}$ [1/K]. This energy scale will be used later for the interpretation of the temperature runs of the instrument. It has to be redefined after each modification of the experimental set-up.

The limits of the possible energy transfers in the instrument are determined by the maximum temperature difference between monochromator and analyzer. However, (4.2) is only valid if $\alpha(T)$ is constant in temperature. For small temperature changes on the order of a few degrees Kelvin this is a good approximation, normally. On a larger scale, the thermal expansion coefficient of Si (Fig.4.11) shows a strong temperature dependence, as seen by Okada and Tokumaru (1984). These authors collected data from the literature and performed their own precision measurements. They concluded that the thermal expansion coefficient of Si follows an empirical formula in the range of 120 K to 1500 K:

$$\alpha(T) = \left(c_1 \left\{1 - e^{-c_2(T-c_3)}\right\} + c_4 \cdot T\right) \cdot 10^{-6} \; [1/\mathrm{K}] \,, \tag{4.3}$$

with the constants $c_1 = 3.725$, $c_2 = 5.88 \cdot 10^{-3}$, $c_3 = 124$ and $c_4 = 5.548 \cdot 10^{-4}$. The curve shown in Fig. 4.11 was calculated according to (4.3).

The precision measurements of Okada and Tokumura (1984) were performed using Bragg reflections of type $(h\,h\,h)$, whereas the reflection type $(h\,h\,0)$ was used by Becker, Seyfried and Siegert (1982). The results of both experiments are in excellent agreement. Therefore, the cubic unit cell is expected to be uniform and no anisotropy factor has to be considered. This is also supported by further investigations of the silicon lattice parameter by Windisch and Becker (1990).

In inelastic experiments dealing with larger temperature differences, the energy transfer is given by

$$E = \frac{a\left(T_{\text{ana}}\right) - a\left(T_{\text{mono}}\right)}{a\left(T_{\text{mono}}\right)} \cdot E_{\text{i}} \, , \tag{4.4}$$

with the lattice parameter $a(T)$ defined as

$$a(T) = a_0 \cdot \left[\int_{T_0}^{T+c_0} \alpha(T') \, dT' + 1\right] \, . \tag{4.5}$$

$a_0 = 5.430741$ Å is the lattice parameter of silicon at $T_0 = 298.2$ K.

Figure 4.12a illustrates the achievable energy transfers by variation of the analyzer temperature at a monochromator temperature of 30 degrees Celsius and different photon energies E_{i} between 13.8 keV and 23.7 keV, due to the corresponding $(h\,h\,h)$ reflections of silicon.

In the present design of the analyzer, temperatures up to 170° C can be selected. Working with a photon energy of 15.8 keV, this corresponds to an energy transfer of about 8 eV. Lowering the temperature of the monochromator crystal and additional cooling to low temperatures (123 K) allow a further increase in the range of energy transfers. This is demonstrated in Fig. 4.12b, showing the energy transfer as a function of the analyzer temperature for different monochromator temperatures and a photon energy of $E_{\text{i}} = 15.8$ keV.

4.2.3 Energy Resolution

For a test of the energy resolution of INELAX, an amorphous sample should scatter photons over the solid angle covered by the area of the analyzer crystal. Therefore, the intensity scattered elastically from fused silica is used to optimize the energy resolution.

The intensity scattered from fused silica as shown in Fig. 4.10, was recorded by using an energy $E_{\text{i}} = 17.8$ keV. The Lorentzian fit shows a halfwidth of 0.2 K corresponding to $\delta E = (9.3 \pm 1)$ meV due to the thermal expansion coefficient of Si at room temperature.

This energy resolution represents the best resolution value obtained so far by INELAX. It can be compared with theoretical calculations of the energy resolution according to Chap. 2. Table 4.2 gives an overview of the different contributions to the energy resolution. For the calculations, most geometrical parameters are taken from Table 4.1. However, the following sizes were optimized for best resolution: beam size at monochromator $D_{\text{ver}} = 1$ mm, $D_{\text{hor}} = 30$ mm; slit VIII $b_{\text{ver}} = 0.6$ mm, $b_{\text{hor}} = 2$ mm; analyzer $D'_{\text{ver}} = 30$ mm, $D'_{\text{hor}} = 76$mm; slit X $b'_{\text{ver}} = 0.8$ mm.

The experimentally observed energy resolution is in excellent agreement with the expected theoretical resolution.

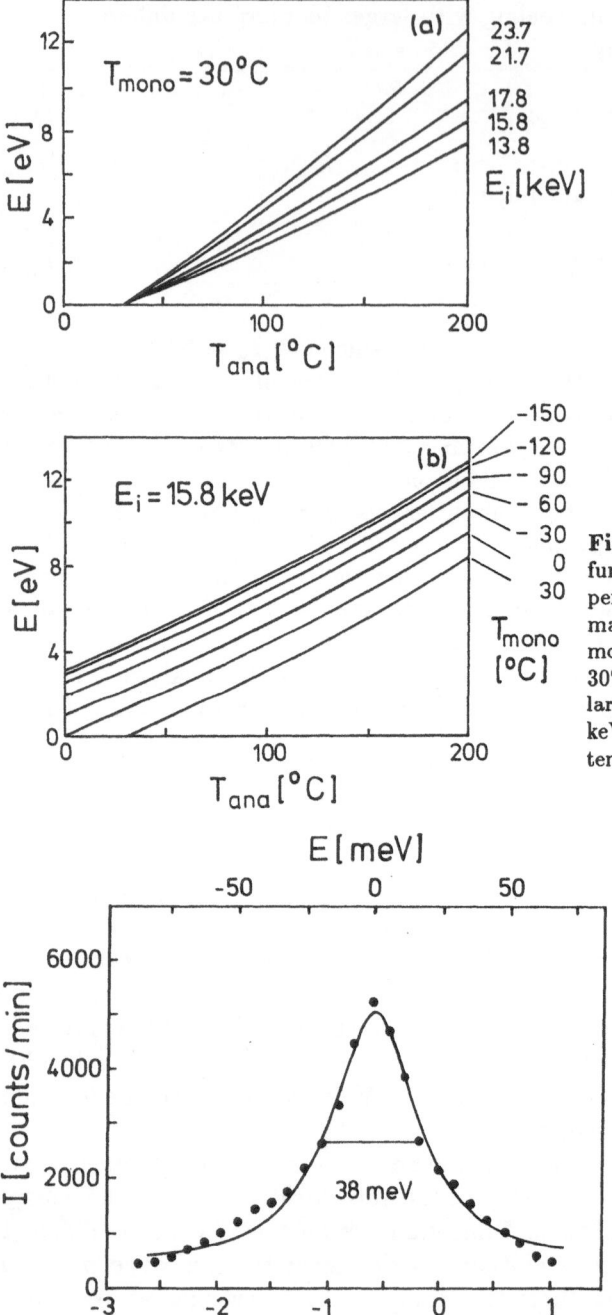

Fig. 4.12. Energy transfer as function of the analyzer temperature (a) for different primary photon energies E_i and a monochromator temperature of 30° C and (b) for the particular photon energy E_i = 15.8 keV at different monochromator temperatures (from Illini 1991)

Fig. 4.13. Intensity elastically scattered from fused silica as a function of the temperature difference between monochromator and analyzer and the corresponding energy transfer. Various slits were opened to obtain higher intensity

Table 4.2. Theoretical contributions to the energy resolution calculated for the (9 9 9) reflection of Si, using the geometrical parameters given in Tab. 4.1 and in the text

	ε_0 [mrad]	$\delta\bar{\varepsilon}$ [mrad]	$(\delta k/k)$ $[10^{-6}]$	δE [meV]
Monochromator				
Vertical geometry	0.375	0.03	0.02	
Horizontal geometry	0	0.74	0.28	
Crystal intrinsic			0.11	
Total			0.30	5.3
Analyzer				
Vertical geometry	0.800	0.25	0.40	
Horizontal geometry	0	0.51	0.13	
Crystal intrinsic			0.11	
Total			0.44	7.8
Spectrometer, total			0.53	9.4

For further improvements of this energy resolution the analyzer geometry has to be optimized according to Table 4.2, where a strong contribution to the energy resolution in the vertical scattering plane can be identified.

For applications with no need for such extreme energy resolution the photon flux at the sample position can be increased by broadening the energy distribution. This can be done by opening the different slit systems. Figure 4.13 shows the intensity elastically scattered from fused silica as a function of temperature difference between monochromator and analyzer for such a case. The energy resolution corresponding to the halfwidth is about 38 meV with an intensity gain of two orders of magnitude. The Lorentzian fit clearly reveals an additional intensity distribution at lower temperature. The origin of this intensity lies in the grooves of the analyzer, where the lattice spacing is a bit enlarged by strains. This demonstrates that the absorbing mixture of glycerine and Ta-powder is not perfectly filling the grooves.

As pointed out in Sect. 4.2.2, the energy transfer of the instrument can reach values of up to about 10 eV. For investigations in this energy range, as in measurements of electronic excitations, there is certainly no necessity for an extremely high energy resolution. In this case the instrumental resolution can be easily accomodated by appropriate slit setting, thus allowing larger beam divergences. This will also increase the photon flux at the sample position.

Further gain of the photon flux at the expense of energy resolution is even possible by changing the focusing crystal elements to ungrooved ones.

Table 4.3. Instrumental details of the INELAX spectrometer

Premonochromator:	double crystal arrangement Si (1 1 1) or Ge (2 2 0)
Monochromator:	Si (h h h), disc of 10 cm diameter, grooved and spherically bent, glued in a glass form or variable curvature
Analyzer:	Si (h h h), disc of 10 cm diameter, grooved and spherically bent, glued in a glass form
Detector:	Si diode at 5° C
Beam size at sample position:	$\sim$ 2 mm x 1 mm
Maximum flux at sample position:	$\sim 2 \cdot 10^7$ photons $\cdot$ mm$^{-2} \cdot$ s^{-1}
Background:	$\leq$ 1 count/min
Incident energy:	13.8 keV, 15.8 keV or 17.8 keV
Range of scattering angle at sample:	$0.5° \leq 2\theta \leq 110°$
Momentum transfer:	0.06 Å$^{-1} \leq Q \leq 14.7$ Å^{-1}
Momentum resolution:	0.03 Å$^{-1} \leq Q \leq 0.3$ Å^{-1}
Energy transfer:	-1 eV $\leq E \leq 10$ eV
Energy resolution:	9 meV $\leq \delta E \leq 40$ meV

4.3 Experimental Parameters of INELAX

Table 4.3 contains an overview of the present capabilities of the INELAX instrument INELAX. Besides showing the already discussed parameters, it also gives the possible range of momentum transfers up to 14.7 Å^{-1} and the achievable momentum resolution.

The photon flux at the sample is also given with $\sim 2 \cdot 10^7$ photons $\cdot$ mm$^{-2} \cdot$ s^{-1}. This value can be compared with the primary flux at the wiggler line of about $3 \cdot 10^{10}$ photons per second and per energy bandwidth of 10 meV at an energy of 18 keV and with the storage ring DORIS working at 3.7 GeV and 100 mA. The theoretical reflectivities of the silicon crystals of the premonochromator and of the monochromator, together with the ratios of crystal acceptance angles and beam divergences and the losses due to absorption in the beam windows (beryllium and kapton foils) reduce the available photon flux by about two orders of magnitude. Besides, the geometrical focusing of the monochromator and the analyzer is also not perfect. Further improvements of these components will decrease the intensity losses.

5. Other Approaches
to Inelastic X-ray Scattering

5.1 Activities Using a Conventional Source

The aim of the group of Egger, Hofmann and Kalus (1984) at Bayreuth University is to develop an experimental set-up for inelastic X-ray scattering at an X-ray generator with a rotating anode, as already mentioned in Sect. 1.3. The promising progress of these activities is reported in this chapter.

The basic difficulty for the performance of a backscattering experiment at an X-ray generator lies in the fact that there exists no suitable combination of the wavelength of a characteristic line and the lattice spacings of monochromator and analyzer crystals at room temperature. However, different combinations of anode materials and crystals can be analyzed and the deviations can be calculated. Because of the desired high energy resolution, only perfect crystals such as silicon and germanium crystals can be taken into consideration. Table 5.1 shows a selection of posssible combinations for a scattering angle of 89.6°. The corresponding mismatch factor A was calculated

Table 5.1. Mismatch factor A between lattice spacing and characteristic wavelength of X-rays for reflections of Si and Ge crystals for a Bragg angle of 89.6°, as calculated by Hofmann (1989)

Material	Reflection $(h\,k\,l)$	Radiation	Mismatch factor A $[10^{-4}]$
Si	(5 3 3)	Ni-K_{α_1}	+ 9.08
Si	(7 5 1)	Ge-K_{α_1}	− 1.24
Si	(14 6 0)	Mo-K_{α_1}	+ 6.72
Si	(15 3 1)	Mo-K_{α_2}	+10.67
Si	(18 4 2)	Pd-K_{α_1}	− 3.07
Si	(17 7 1)	Pd-K_{α_2}	− 1.68
Si	(17 9 1)	Ag-K_{α_1}	− 2.08
Ge	(6 2 0)	Co-K_{α_1}	− 0.35
Ge	(15 1 1)	Nb-K_{α_2}	− 7.32
Ge	(15 5 1)	Mo-K_{α_2}	− 8.32
Ge	(16 8 4)	Rh-K_{α_2}	+ 5.78

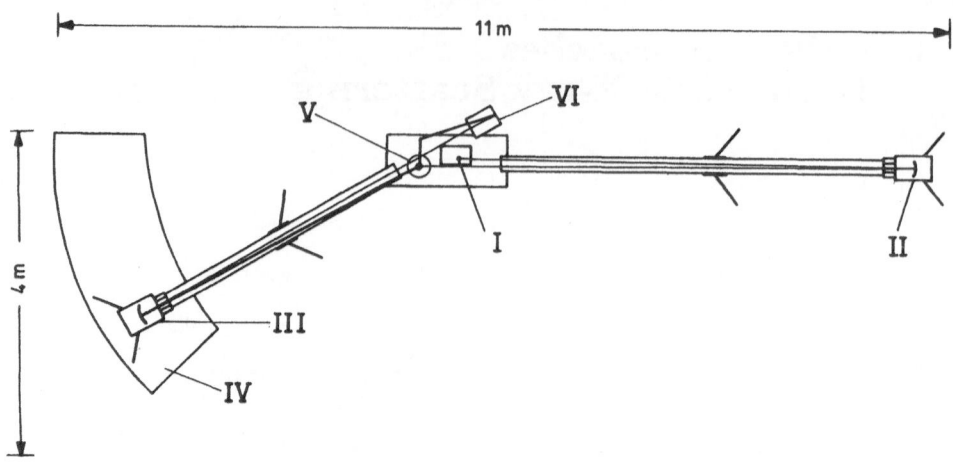

Fig. 5.1. Schematic view of the Bayreuth spectrometer at an X-ray generator (I) with a rotating anode (Hofmann 1989). The beam illuminating the monochromator (II) is backscattered onto the sample (V). The scattered intensity is then focused by the analyzer (III), which is aligned on a granite baseplate (IV), into the detector (VI)

by Hofmann (1989), using

$$A = \frac{\lambda_{K_\alpha}}{2d_{hkl} \cdot \sin\theta} - 1 \tag{5.1}$$

with d_{hkl} describing the lattice spacing. λ_{K_α} is the wavelength of the characteristic X-ray line.

A certain combination of crystal and radiation is suitable for a backscattering experiment, if the according mismatch factor A in Table 5.1 can be reduced to zero. A reduction of the scattering angle will diminish the factor A in case of small negative mismatch values. However, this is not true for positive mismatch values. In these cases, the thermal expansion of the crystals is suitable to obtain the optimal match by the selection of the appropriate temperature T_{opt}. This temperature value is determined by

$$\int_{T_0}^{T_{opt}} \alpha(T)\, dT - A = 0. \tag{5.2}$$

The thermal expansion coefficient of Si is given by (4.3). The Bayreuth group decided to work with a nickel anode and, therefore, selected the (5 3 3) reflection of silicon. According to (5.2), the corresponding temperature for optimal match is 296.1° C. The intrinsic energy resolution for silicon is given by (2.20) and Fig. 2.5 and is 30.5 meV. According to (2.25), additional contributions due to beam geometry have to be added to obtain the total energy resolution of the basic components.

The design of the Bayreuth spectrometer at the X-ray generator (I) is shown in Fig. 5.1 (Hofmann 1989). The fundamental scheme is equivalent to

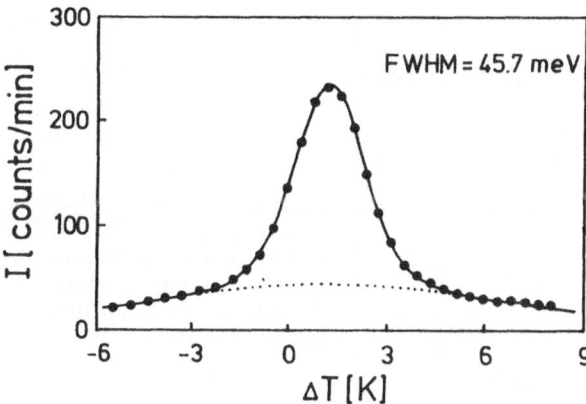

Fig. 5.2. Scattered intensity from pyrolytic graphite (0 0 2) as a function of the temperature difference between analyzer and monochromator. The drawn *lines* are fits to the data (from Hofmann 1989)

Fig. 2.4 and Fig. 4.1. In contrast to the INELAX instrument at the storage ring at DESY, this instrument works in the horizontal scattering plane. The monochromator (II) and the analyzer (III) crystals are both spherically bent. They are mounted in special housings which allow heating up to the necessary temperatures and which are installed at the ends of about 5 meter long vacuum tubes. The analyzer arm moves on a polished granite baseplate (IV). The sample position (V) and the arrangement of the detector (VI) are shown as well. The variation of the energy transfer at this instrument is done by varying the temperature of the analyzer crystal in analogy to the INELAX technique as already described in Sect. 4.2.

In this particular geometrical arrangement the overall energy resolution of the instrument was projected to be $\delta E = 51$ meV. Test measurements for the determination of the energy resolution were performed by using pyrolytic graphite. Figure 5.2 shows the scattered intensity for the (0 0 2) Bragg reflection of graphite as a function of the temperature difference between analyzer and monochromator. The energy halfwidth was determined to be 45.7 meV. The broad background distribution is attributed to an intensity originating from the distorted parts of the faceted monochromator crystal. Yet the Bragg reflection of graphite is not completely illuminating the analyzer area, and is therefore not a true indicator of the total instrumental resolution. An improved test was performed by resolving energetically a Debye-Scherrer ring of graphite powder. The observed scattering signal is shown in Fig. 5.3 and reveals an energy resolution of about 66 meV. With the successful performance of the instrument at the rotating anode generator first attempts for inelastic measurements of phonon excitations are projected. The maximum possible ideal photon flux on the sample in this arrangement is $7 \cdot 10^6$ photons $\cdot$ mm$^{-2} \cdot$ s^{-1}, calculated with a theoretical energy resolution of 51 meV (Hofmann 1989). This value of the photon flux lies just below the presently achieved flux at INELAX (Table 4.3).

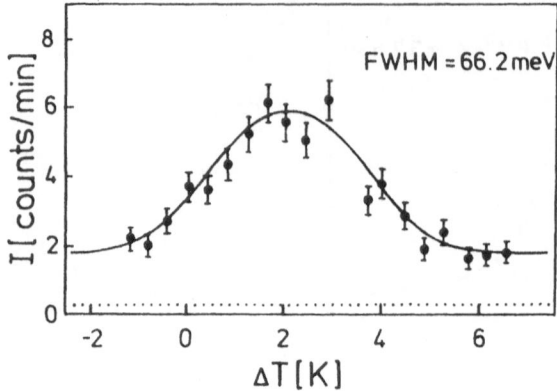

Fig. 5.3. Scattered intensity from graphite powder (0 0 2) as a function of the temperature difference between analyzer and monochromator (from Hofmann 1989)

An important improvement can be achieved by the development of alloy crystals of the type Si_xGe_{1-x} (Burkel and Peisl 1985). For these systems any lattice parameter between the values for silicon and germanium can be obtained. This would allow a very elegant method of reducing the mismatch factor A for a certain characteristic wavelength and for avoiding extreme heating of the analyzer. As a realistic combination it is proposed to use the (9 9 9) reflection of a $Si_{0.568}Ge_{0.432}$ alloy crystal together with the characteristic K_{α_1} line of a molybdenum anode. However, the perfection of silicon crystals with such high Ge concentrations is not yet sufficient to obtain the desired energy resolution (Magerl, Heidemann, Holm and Sirtl 1984).

5.2 Activities Using an Undulator

The interest in inelastic X-ray scattering with high energy resolution also motivated activity at the Brookhaven National Laboratory in the United States (Sect. 1.3). Test experiments using the X-ray undulator source of the PEP storage ring were performed at SSRL at Stanford.

The instrumental set-up used in this experiment is shown schematically in Fig. 5.4 (Siddons, Hastings, Moncton, Hewitt and Brown 1986). In this arrangement the beam path from source (I) to premonochromator (II) and monochromator (III) is very similar to INELAX. Therefore, it is certainly sufficient to focus only on the modifications. The intensity scattered by the sample (IV) is energetically analyzed by a cylindrically bent and grooved silicon crystal (V) with sagittal focusing. The radiation is then dispersed according to its energy along a one-dimensional position sensitive detector (VI) at the focal position of the analyzer crystal. The intensity in the spectrum obtained by this method is a function of the energy transfer of the photons.

The energy resolution of this set-up as measured by the elastically scattered intensity from a Plexiglas sample is 37 $\pm$ 2 meV. Due to the intense

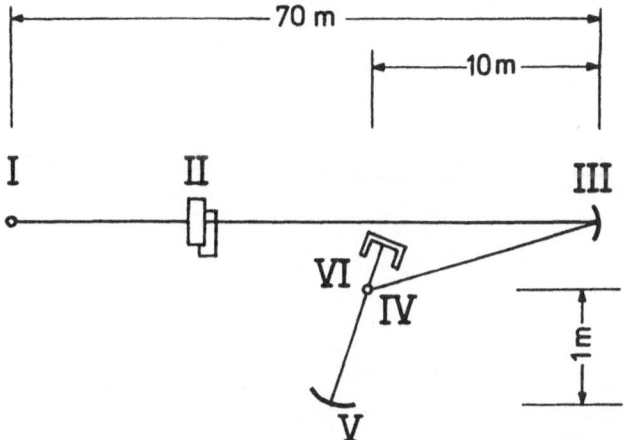

Fig. 5.4. Schematic view of the instrumental layout used for the experiment at the PEP undulator, as taken from Siddons, Hastings, Moncton, Hewitt and Brown (1986). The beam from the source (I) is passing the premonochromator (II) and the monochromator (III) on its way to the sample (IV). The scattered intensity is focused into the detector (VI)

undulator source, the measured intensity at the sample position was greater than 10^9 photons/s. Despite this high intensity there was no inelastic X-ray scattering observed at that time.

In a recent development, the group at the Brookhaven National Laboratory used a modified crystal combination (Siddons, Hastings and Faigel 1988) for a monochromator in a measurement of nuclear Bragg scattering (Hastings, Siddons, van Bürck, Hollatz and Bergmann 1990). The schematic view of this arrangement is shown in Fig. 5.5. The energy bandwidth of the X-ray beam passing the premonochromator [(I) with Si (1 1 1)] is further reduced by two dispersive reflections (II) of the type Si (10 6 4). An energy resolution of 5 meV at 14.4 keV photon energy was achieved.

This interesting combination allows scattering in forward direction. Therefore, it is certainly well suited as a basis for an inelastic scattering experiment, especially in new-generation INELAX instruments at an undulator source, because of the small beam size and of the small beam divergence (see Sect. 7.2).

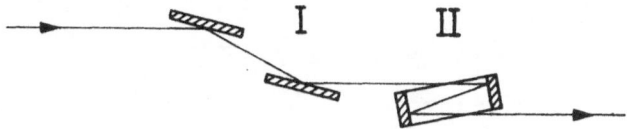

Fig. 5.5. Crystal arrangement to obtain high energy resolution with scattering in the forward direction. From Hastings et al (1990). The beam is reflected by the premonochromator (I) and and another dispersive two crystal arrangement (II)

6. Applications of Inelastic X-ray Scattering

6.1 Phonons in Single Crystals

Inelastic scattering of X-rays from phonons is expected to supply similar information about the dynamics of the observed system as coherent inelastic neutron scattering does. This is supported by the already discussed experiments of Joynson (1954) and Walker (1956) (see Sect. 1.1). Within the limits of the adiabatic approximation, the electrons are expected to follow the movements of the nuclei instantaneously. Therefore, phonons, i. e., low frequency motions of the nuclei, will cause electron charge density variations, which can be directly observed by inelastic X-ray scattering.

The scattering function for one phonon scattering can be written as

$$S(\boldsymbol{Q}, \omega) = G(\boldsymbol{Q}, \boldsymbol{q}, j) \cdot F(\omega, T, \boldsymbol{q}, j). \tag{6.1}$$

The dynamical structure factor $G(\boldsymbol{Q}, \boldsymbol{q}, j)$ is given by

$$G(\boldsymbol{Q}, \boldsymbol{q}, j) = \left| \sum_{d}^{\text{unit cell}} f_d(Q) \cdot e^{-W_d} \left[\boldsymbol{Q} \cdot \boldsymbol{e}_d(\boldsymbol{q}, j) \right] M_d^{-1/2} e^{i\boldsymbol{Q}\boldsymbol{d}} \right|^2. \tag{6.2}$$

$f_d(Q)$ is the atomic form factor of atom d at position $\boldsymbol{d}$ with $f_d(Q \rightarrow 0) = Z$. Z is the atomic number. $\boldsymbol{e}_d(\boldsymbol{q}, j)$ is the component of the normalized phonon eigenvector in the mode j with phonon wavevector $\boldsymbol{q}$ for atom d. e^{-W_d} is the Debye-Waller factor of atom d and M_d is its mass. The function $F(\omega, T, \boldsymbol{q}, j)$ for undamped phonons is given by

$$F(\omega, T, \boldsymbol{q}, j) = \frac{\langle n \rangle + 1/2 \pm 1/2}{\omega_{\boldsymbol{q}, j}} \cdot \delta \left(\omega \mp \omega_{\boldsymbol{q}, j} \right). \tag{6.3}$$

Here the upper sign holds for energy loss and the lower one for energy gain by the X-rays. $\langle n \rangle$ is the Bose occupation factor and $\omega_{\boldsymbol{q}, j}$ the frequency of phonon mode j with wavevector $\boldsymbol{q}$.

The scattering law for X-rays is obtained from the scattering law for neutrons by simply replacing the Fermi scattering length by the atomic form factor.

Detailed discussions of the theory of phonons are given by Peierls (1955), Cochran and Cowley (1967) or Horton and Maradudin (1974).

The phonon eigenvectors $\boldsymbol{e}(\boldsymbol{q}, j)$ define the direction of the atomic vibrations in the planes perpendicular to the direction of the plane wave given

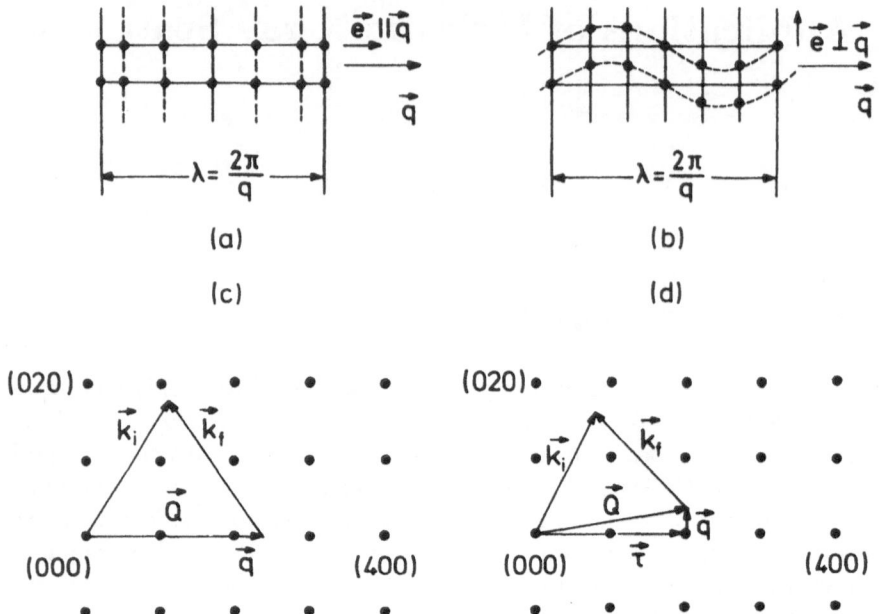

Fig. 6.1. Atomic vibrations in the main symmetry direction of a crystal with longitudinal (a) and transverse (b) polarization in real space. Scattering geometry in reciprocal space for the measurement of a longitudinal (c) and transverse phonon (d) in [1 0 0] direction

by q. In the main symmetry axes of crystals two different vibrational modes exist, Fig. 6.1a, b: a longitudinal mode which is connected with atomic displacements parallel to the direction of the wavevector ($e \parallel q$), and a transverse wave which is connected with atomic displacements perpendicular to the direction of the wavevector ($e \perp q$). The eigenvectors are therefore also called polarization vectors. In general directions mixed vibrational modes are present.

The scattering geometry for detecting phonons in crystals is illustrated in Fig. 6.1c, d. It shows the positions of the scattering vectors in reciprocal space [(h k 0) plane] for the detection of longitudinal (Fig. 6.1a) and transverse (Fig. 6.1b) phonons in the [1 0 0] direction. The scattering vector Q can be expressed by (see 2.2)

$$Q = k_i - k_f = \tau + q \,, \tag{6.4}$$

where q is the phonon wavevector, defined with respect to the nearest reciprocal lattice point τ. The directions of the polarization vector $e(q)$ and the scattering vector Q select the type of vibrational mode to be detected, according to (6.2). An energy scan with such a scattering vector will show two signals, due to the creation and the annihilation of a phonon.

The first successful attempts for inelastic, energy resolved X-ray scattering were performed by Burkel, Peisl and Dorner (1987) with pyrolytic graphite and beryllium. These materials were chosen because they have high energetic optic modes.

6.1.1 Beryllium

For the measurement of a phonon dispersion curve by inelastic X-ray scattering, hcp beryllium shows some advantages. The absorption of hard X-rays is very small. Single high quality crystals are available.

The longitudinal (L) modes in the [0 0 ξ] direction are particularly suited for investigations in Brillouin zones of (0 0 l) type. The structure factor is either zero or $4 \cdot f^2(Q) \cdot Q^2$, depending on l and on whether the mode is acoustic or optic. This is due to the fact that Bragg intensities at (0 0 l) in an hcp structure with two atoms per primitive unit cell are extinguished for odd l and are proportional to $4 \cdot f^2(Q)$ for even l. Therefore, the longitudinal acoustic

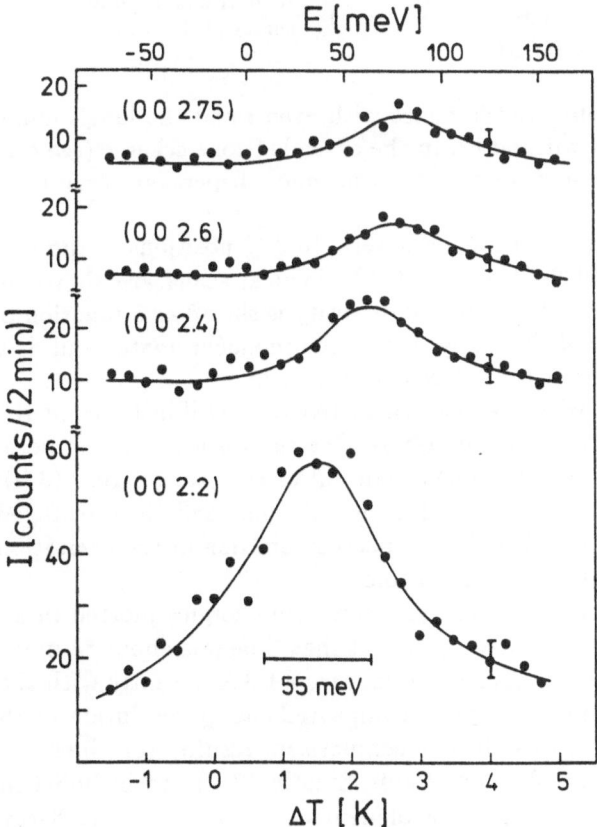

Fig. 6.2. Scatttered X-ray intensity for constant-Q scans at different Q values for longitudinal modes in the [0 0 ξ] direction for Be

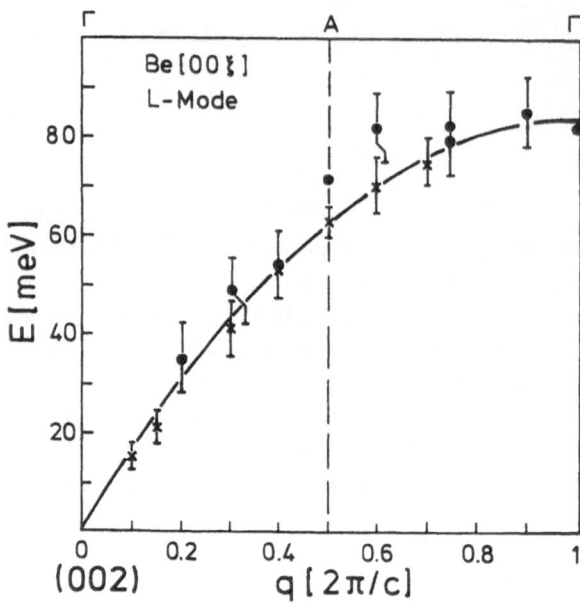

Fig. 6.3. Dispersion of longitudinal modes in Be in the [0 0 ξ] direction. The X-ray results (● 1987, × 1989) are compared with results from neutron scattering (— Stedman, Amilius, Pauli, Sundin 1976 with an experimantal accuracy of about line thickness)

(LA) modes can only be observed in zones with even l and the longitudinal optic (LO) modes in zones with odd l. In the extended zone scheme (see also Fig. 6.3), the longitudinal modes present a continous dispersion curve from one Γ-point to the next Γ-point.

Constant-Q scans were performed at several (0 0 ξ) positions in the first Brillouin zones (0 0 1), (0 0 2) and (0 0 3). Typical scans are shown in Fig. 6.2 for the (0 0 2) zone. The scattered intensity is shown as a function of the temperature difference of the analyzer and the monochromator and as a function of the energy transfer for different values of Q.

These data were taken with the instrument INELAX still installed at the bending magnet of the storage ring DORIS II. One run through the temperature range took as long as the lifetime of one fill of the storage ring (3 h). All the data were therefore accumulated in several runs, and then averaged for temperature intervals of 0.25 K. The expected variation of the excitation energies of the phonon peaks is clearly visible.

Figure 6.3 shows the observed phonon excitation energies plotted in an extended zone scheme. The energy resolution at that time was about 55 meV. Because of the large error bars, additional data were taken recently with the instrument at the wiggler line and with an improved energy resolution of 18 meV. These data are shown as well. For comparison, results from inelastic neutron scattering (Stedman, Amilius, Pauli, Sundin 1976) are included in the same figure. The direct comparison of neutron with the recent X-ray scattering results shows excellent agreement.

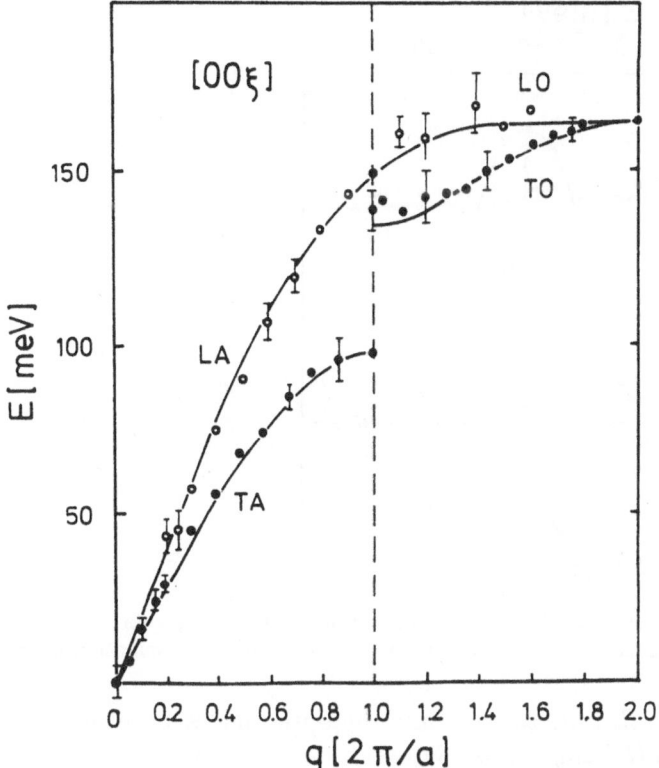

Fig. 6.4. Phonon dispersion curves for longitudinal (o) and transverse (•) modes in the [0 0 ξ] direction in diamond. The results from X-ray scattering are shown together with a shell-model fit from the literature (Warren, Yarnell, Dolling and Cowley 1967) in an extended zone scheme

6.1.2 Diamond

Diamond is an excellent species on which to demonstrate the capabilities of inelastic X-ray scattering, because a small sample of only several mm^3 can be used. Due to the very high frequencies of the optical phonon modes, conventional neutron scattering at a reactor source is tedious and thus, there are not many experimental data available for the optic modes.

For the X-ray investigations, symmetry directions with "pure" acoustic and optic modes, either longitudinal or transverse, were chosen again. The scattering geometry was chosen so as to reach an (almost) vanishing dynamical structure factor, see (6.2), for one kind of polarization.

Diamond has an fcc Bravais lattice with two atoms per primitive unit cell. Consequently acoustic and optic modes appear and are degenerate at the zone boundary in the [ξ 0 0] direction. The two branches merge into the zone boundary with finite slopes, opposite in sign. For the experiment these longitudinal modes have the advantage that the acoustic modes are

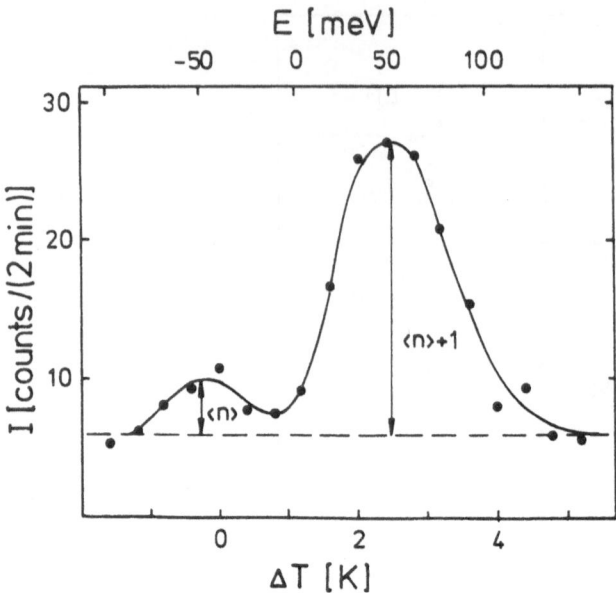

Fig. 6.5. The energy scan at $Q = (0\ 0\ 4.2)$ in diamond shows the energy gain and energy loss signal of the longitudinal phonon clearly separated. $\langle n \rangle$ is the Bose occupation factor

exclusively visible in $(4h\ 0\ 0)$ zones, while the optic modes have intensity only in zones $(2h\ 0\ 0)$ (h being an odd integer).

The experiments were performed between the reciprocal points $(0\ 0\ 4)$ and $(0\ 0\ 6)$ for the longitudinal and between $(0\ 0\ 4)$ and $(0\ 1.8\ 4)$ for the transverse phonon modes. The results are shown together in the extended zone scheme in Figure 6.4. The drawn line is taken from the literature and shows a shell-model fit to data from neutron scattering (Warren, Yarnell, Dolling and Cowley 1967). The good agreement of the X-ray data with this fit is a direct proof that both methods, X-ray and neutron scattering, lead to the same frequencies of the phonons.

An energy scan at $(0\ 0\ 4.2)$ is given in Fig. 6.5. It shows very distinctly both phonon signals, for energy gain and energy loss. By using the measured phonon energy and the sample temperature (300 K), the Bose occupation factor $\langle n \rangle$ is calculated. With this factor the intensity ratio of the phonon signals can be calculated according to the scattering law. This ratio is in excellent agreement with the ratio of the peak intensities, as demonstrated in Fig. 6.5.

The transverse modes in the $[0\ 0\ \xi]$ direction in diamond are doubly degenerate. No additional degeneracy appears at the zone boundary and both the acoustic and optic branches cause scattering intensity near the zone boundary. Figure 6.6 shows that the energy gap between the two modes is already resolved by the moderate energy resolution of about $\delta E = 19$ meV, used in this case.

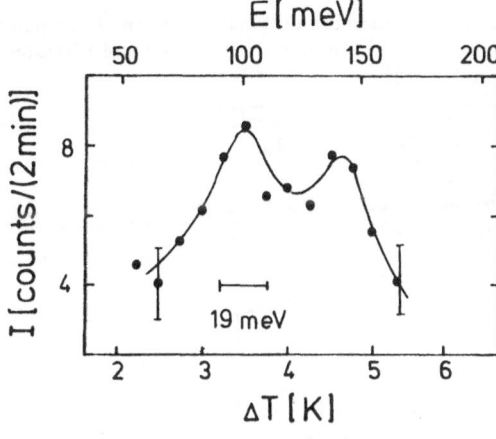

Fig. 6.6. The energy scan at $Q =$ (0 1 4.2) in diamond shows the gap between the transverse acoustic and the transverse optic mode. The *line* is drawn only to guide the eye

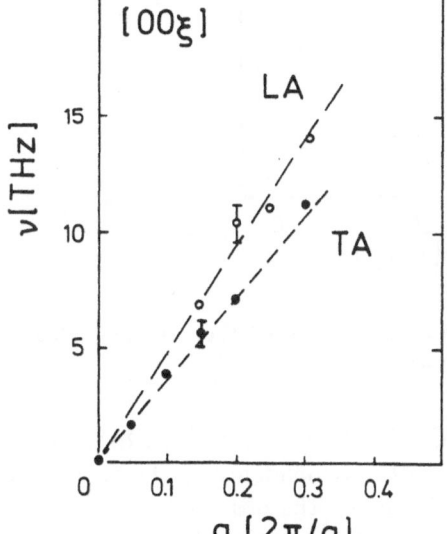

Fig. 6.7. Initial slopes of the dispersion curves in diamond for the longitudinal and transverse acoustic modes in the [0 0 ξ] direction

The elastic constants of diamond can also be determined by means of inelastic X-ray scattering. Figure 6.7 shows the initial slopes of the dispersion curves in the longitudinal and transverse direction in the [0 0 ξ] zone. The values derived for the elastic constants c_{11} and c_{44} from these slopes are compared with values known from Brillouin scattering (Grimsditch and Ramdas 1975) and from ultrasonic pulse experiments (McSkimin and Andreatch 1972) in Table 6.1. There is good agreement between the observations in Fig. 6.7 and the literature.

A further test of the validity of the scattering function for one-phonon scattering is to analyze the intensities of the observed phonon excitations according to (6.1) and (6.2). The energy resolution for the experimental constant-Q scans is a constant (Sect. 2.4), therefore, the phonon intensity is correctly given by the peak intensity. According to the scattering function

Table 6.1. X-ray results for elastic constants of diamond in units of $[10^6 \text{ N/cm}^2]$ compared to values from Brillouin scattering (Grimsditch and Ramdas 1975) and from ultrasonic pulse experiments (McSkimin and Andreatch 1972)

Phonon mode	Elastic constant	Inelastic X-ray scattering	Brillouin scattering	Ultrasonic pulse
TA	c_{44}	57.4 ± 3.0	57.74 ± 0.14	57.8 ± 0.2
LA	c_{11}	99.1 ± 9.0	107.6 ± 0.02	107.9 ± 0.5

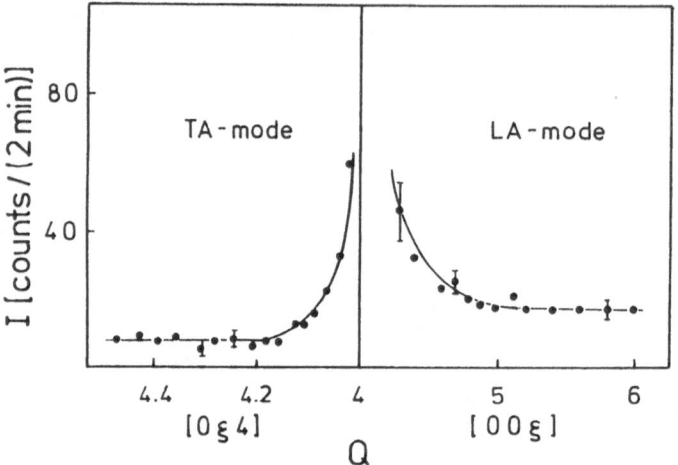

Fig. 6.8. Scattering intensity of the TA and LA phonons in the $[0\ 0\ \xi]$ direction for different Q values, compared with the calculated (—) intensity

the intensity is proportional to $Q^2 \cdot f^2(Q) \cdot e^{-2W} \cdot (\langle n \rangle + 1)/\omega_j$. In addition to this, a factor $(\sin 2\theta)^{-1}$ appears due to the dependence of the scattering volume on the scattering angle. Figure 6.8 shows the observed intensities at different Q values together with a curve for the calculated values at $Q = (0\ 0\ 5.8)$ for a longitudinal and a transverse mode in diamond. The good agreement for different momentum and energy transfer proves that the experimental data are well described by the scattering function for inelastic X-ray scattering.

6.1.3 High Temperature Superconductor

One of the advantages of X-ray scattering is the possibility of investigating the scattering from very small samples. Certainly the new generation of high temperature superconductors are systems which are at present only available in the form of small single crystals. Knowledge about the lattice dynamics in these systems will provide very important information on the interaction

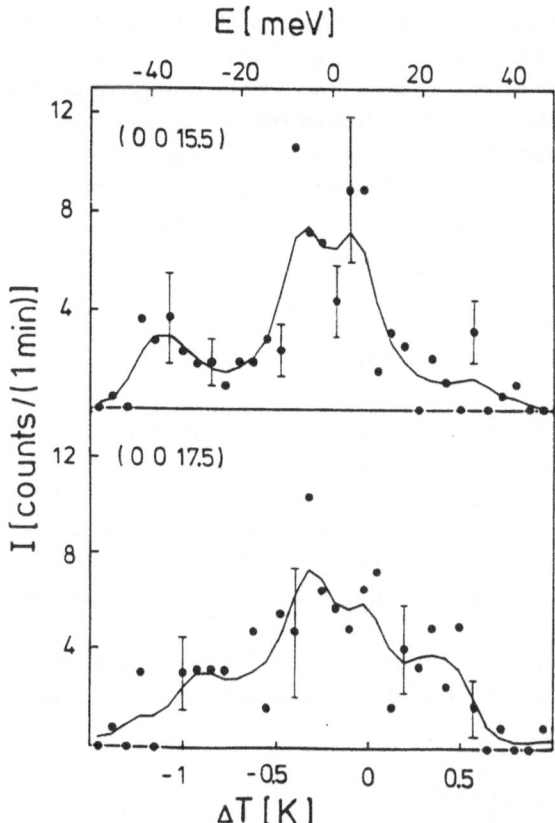

Fig. 6.9. X-ray intensity scattered from $Tl_2Ba_2CaCu_2O_8$ for constant-Q scans at (0 0 15.5) and (0 0 17.5). The *lines* are drawn by a smoothing routine

potentials and, thus, lead to further information on the mechanism of superconductivity at high temperatures. Knowing the phonon spectrum will help to investigate the validity of the BCS theory for the superconductivity in these materials. Calculations using a shell model were already performed on new ceramic superconductors with perovskite structure (Prade 1989). Yet, there are only few experimental data available to adjust the mostly assumed interaction potentials. Infrared absorption (Sugai, Ucida, Takagi, Kitazawa and Tanaka 1987; Vuong 1987) and Raman scattering (Brun et al 1987; Blumenroeder 1987) were performed, but results from inelastic neutron scattering are limited (for a review see Pintschovius 1990). Due to the complicated structure of these systems, the phonon dispersion shows a large number of branches.

However, calculations by Strauch (1990) on $Tl_2Ba_2CaCu_2O_8$ revealed some isolated modes, for which the neighboring modes have vanishing structure factors. This observation on $Tl_2Ba_2CaCu_2O_8$ superconductors (provided H. H. Otto and G. Jehl, Regensburg) encouraged first test experiments on single crystals (scattering volume about 0.15 mm³) by means of inelastic X-ray scattering. Figure 6.9 shows the result for constant-Q scans at (0 0 15.5) and (0 0 17.5) together with lines drawn by a smoothing routine. The spectra reveal a certain fine structure and an optimistic interpretation might correlate it with possible gain and loss signals at energy transfers of about 6 and 35 meV and 6 and 25 meV, respectively. These transfers can be attributed to longitudinal phonon modes. The intensity ratios of the energy gain and loss signals cannot be measured conclusively due to poor statistics. Nevertheless, after these encouraging preliminary observations a more detailed investigation of these systems will follow.

6.2 Vibrational Excitations in Non-single-crystal Systems

The presented results for the determination of phonon dispersion curves in the simple single crystals of Be and diamond clearly prove the suitability of inelastic X-ray scattering for such measurements. The investigations of the high temperature superconductor sample are a promising attempt for a challenging application of the method to very small crystals. But also other investigations with inelastic X-ray scattering, complementary to inelastic neutron scattering, are possible.

The relations between wavevector and energy for X-rays and neutrons are different due to the fact that neutrons have, in contrast to X-rays, a mass (here denoted by m):

$$
\begin{aligned}
\text{neutrons}: \hbar\omega &= E_i - E_f = \frac{\hbar^2}{2m}\left(k_i^2 - k_f^2\right) \\
\text{X} - \text{rays}: \hbar\omega &= E_i - E_f = c\hbar\left(k_i - k_f\right) .
\end{aligned}
\tag{6.5}
$$

Since thermal neutrons have energies comparable to excitations in solid matter, their wavevector k_i changes considerably in an inelastic scattering process. Therefore, carrying out constant-Q scans is rather difficult and requires the use of sophisticated instruments, like three axis spectrometers.

In inelastic X-ray scattering the wavevector k_f has approximately the same magnitude as the incoming wavevector k_i. Therefore, every inelastic scan at constant scattering angle θ is automatically a constant-Q scan.

A direct consequence of this is further illustrated in Fig. 6.10, which shows the accessible range of energy and momentum transfers for X-rays and neutrons. The range of small momentum transfer Q in connection with a large energy transfer $\hbar\omega$ is out of reach for neutrons, but it can be obtained by X-ray scattering. The allowed region is a function of the incident neutron energy E_i and lies between the (E, Q)-loci corresponding to forward and backward scattering.

Collective excitations in liquids or in hard polycrystalline materials have a high sound velocity v and, therefore, can be investigated only by means of X-ray scattering. The thermal neutron scattering method is applicable only in the range of velocities smaller than 3000 m/sec (Dorner 1966). In Fig. 6.10

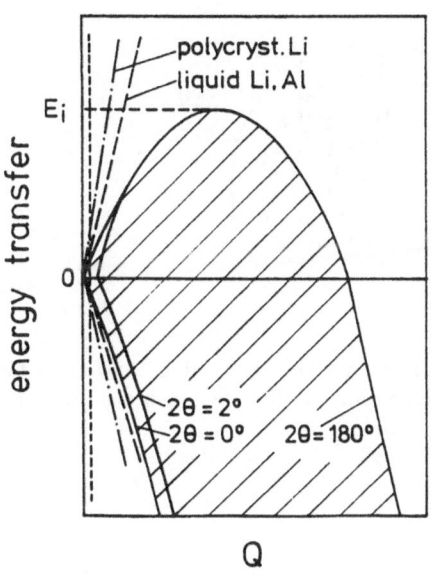

Fig. 6.10. Experimentally accessible range for inelastic scattering of neutrons with the energy E_i. The *broken lines* indicate sound velocities in liquid Li and Al and polycrystalline Li. A constant-Q scan for X-rays is given as *short-dashed* line parallel to the energy axis

the slope of the excitation dispersion is shown for liquid Al and Li ($v_{Al}= 4650$ m/s), for liquid Li ($v_{Li}=4517$ m/s, Ruppersberg and Speicher 1976) and for polycrystalline Li (this work).

6.2.1 Polycrystalline Lithium

The investigations of polycrystalline lithium, which will be presented in this review, should not be expected to be a complete analysis. They were performed as a test for vibrational excitations in non-single-crystalline systems at small momentum transfers during developing the inelastic X-ray scattering techniques (Gauss 1989; Burkel, Gauss, Illini and Peisl 1990).

Polycrystals contain a variety of randomly oriented single crystallites. There is no distinct orientational direction as in single crystals and, therefore, the direction of the momentum transfer $\hbar Q$ is not defined. Because of the averaging over the orientation only the absolute value, $|Q| = Q$, is defined. This means that the scattering process takes place on a spherical shell in the reciprocal space and not on a point, as in the case of single crystals. In an elastic scattering process the shells are observed as Debye-Scherrer rings. This is illustrated in Fig. 6.11 in analogy to Fig. 6.1. There can be fine structure of the rings, depending on the grain size of the crystallites and on the resolution volume.

The attempt to measure phonons with a certain momentum transfer can lead to contributions from a variety of phonon modes, depending on the value of Q and on the solid angle $d\Omega$ (Fig. 6.11). In order to obtain a dispersion relation of a material, one needs a distinct correlation of the wavevector q with its frequency ω. In a polycrystalline material this is only possible in

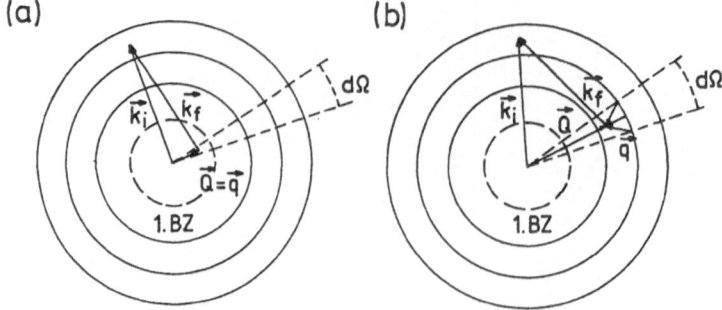

Fig. 6.11. The scattering geometry in a polycrystal for the observation of phonons for momentum transfers **Q** in the reciprocal space within the first Brillouin zone (a) and beyond (b)

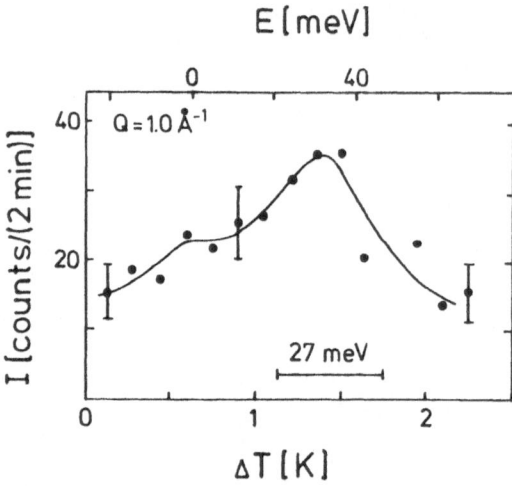

Fig. 6.12. Energy scan for polycrystalline lithium at $Q = 1.0$ Å^{-1}. The *line* is drawn as a guide for the eye

the first Brillouin zone with the reciprocal lattice vector $\boldsymbol{\tau}$ equal to 0, which leads to the fact that the momentum transfer $\hbar Q$ equals $\hbar q$. This shows that only longitudinal (compression) phonon modes are measured within the first Brillouin zone (Fig. 6.11a). Any momentum transfer beyond the first Brillouin zone will excite several longitudinal and transverse phonon modes at a time and the result will be a scattering spectrum which gives information on the density of states of these vibrational modes (Fig. 6.11b).

Lithium was chosen because of its low absorption cross section and its high velocity of sound, which prohibits studying it with neutrons at small momentum transfers, as explained before.

Figure 6.12 shows the scattered X-ray intensity as a function of the temperature difference between the monochromator and the analyzer and the corresponding energy transfer for polycrystalline lithium at room temperature at $Q = 1.0$ Å^{-1}. Despite the large error bars, one can detect the energy loss peak at $E = 33 \pm 5$ meV due to longitudinal vibrations.

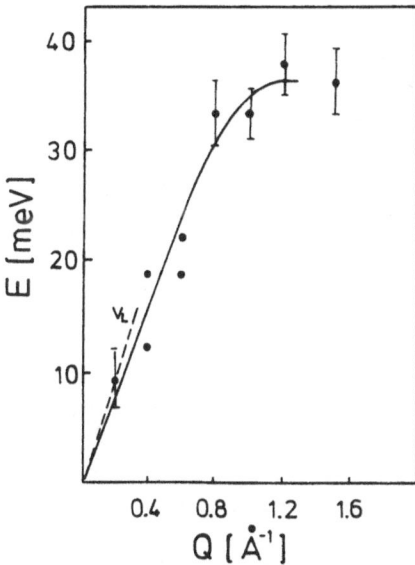

Fig. 6.13. Dispersion relation for longitudinal (compression) modes in polycrystalline lithium. The *line* is drawn to guide the eye

Figure 6.13 gives information on energy transfers gained from the test experiments as a function of the momentum transfer. A dispersion of longitudinal modes for polycrystalline lithum in the first Brillouin zone can be observed. These modes can also be described as compression modes. The longitudinal velocity of sound, taken from the literature (D'Ans-Lax 1967), is also indicated in Fig. 6.13. This value is in agreement with the experimental value within the error bars. A direct comparison of the observed energies with the values obtained from neutron measurements on single crystals in higher Brillouin zones is problematic, since the grain sizes of the polycrystal, which should be small compared to the resolution volume, were not determined.

These data successfully demonstrate the detection of vibrational excitations in the first Brillouin zone at small scattering angles by means of inelastic X-ray scattering. As an application of this technique, amorphous and glassy systems can be studied. Despite the absence of translational symmetry, they have long-wavelength excitations at high energies at small wavevectors.

6.2.2 Pyrolytic Graphite

At this point, a short excursion should be made back to the first application of inelastic X-ray scattering, certainly for completeness of this review, but also because in this case inelastic measurements with X-rays and neutrons were performed on the same sample.

Pyrolytic graphite has high energy optic modes whose amplitudes are purely confined to the hexagonal plane (Al-Jishi, Dresselhaus 1982). Therefore, the hexagonal plane can be selected as the scattering plane. Due to the scattering geometry and the smearing of ($h\ k\ 0$) reciprocal lattice points into

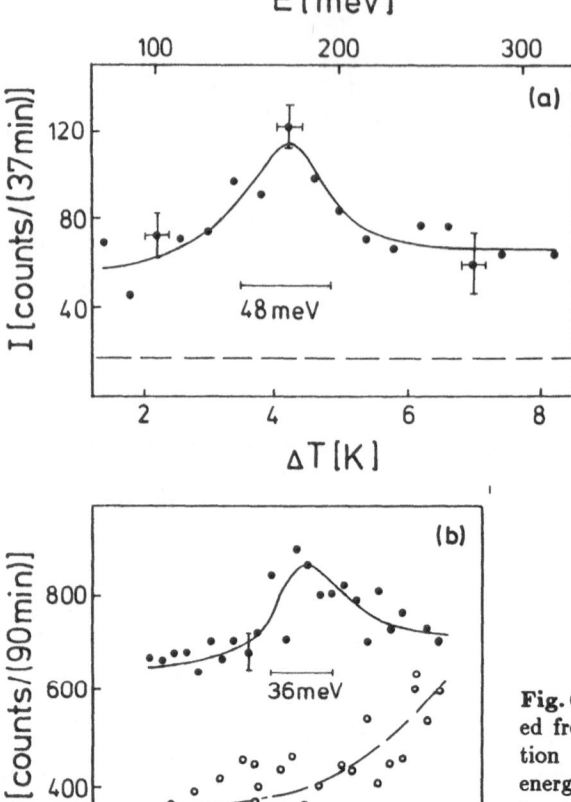

Fig. 6.14a,b. X-ray intensity scattered from pyrolytic graphite as a function of temperature difference and energy transfer (a). Inelastic neutron scattering intensity from pyrolytic graphite as a function of energy transfer in a constant-Q scan (b). In both cases the background is given as well

rings around the hexagonal c-axis, a maximum in the density of states rather than a single phonon excitation is observed, as in the polycrystalline sample discussed before.

Figure 6.14a shows the scattered X-ray intensity at $Q = (0\ 0\ 1.2)$ as a function of the temperature difference and of the correlated energy transfer, a result of 12 individual temperature scans averaged over 0.4 K steps. The pronounced maximum in the scattering appears at an energy transfer of $E = 171 \pm 10$ meV.

To compare X-ray and neutron experiments, the same scans were measured in an equivalent scattering geometry with the IN1 three axis spectrometer of the ILL, Grenoble (Burkel, Peisl and Dorner 1987). Here the results of 10 individual constant-Q scans were added together. The incoming neutron energy had to be varied in order to separate spurions. The final result in Fig. 6.14b shows a maximum at $E = 178 \pm 5$ meV. The calculated instrumental uncertainty is 36 meV. The energy transfer is, within the error bars, the same as for inelastic X-ray scattering. Both results also agree with the calculated density of states for the transverse high-frequency modes of pyrolytic graphite in the hexagonal plane (Al-Jishi, Dresselhaus 1982), which appear at 170 meV with a distribution over 25 meV.

The most impressing comparison is the one concerning the scattering volumes: in the neutron scattering experiment the size of the sample was 6 cm^3, while in the X-ray experiment a sample 200 times smaller was sufficient. This success was already possible at the beginning of the development of the inelastic X-ray scattering technique even with the relatively low intensity provided by a bending magnet.

6.2.3 Liquid Lithium

As indicated in Fig. 6.10, the dynamics of liquids might open an interesting research area for inelastic X-ray scattering, yielding information complementary to that from inelastic neutron scattering.

A phonon in the range of sound propagation is a collective mode in local thermodynamic equilibrium and can be excited by a long wavelength probe. Exciting a liquid with such a probe gives a response which is normally described within the hydrodynamic limit. The density fluctuations which arise are characterized by the statistically independent variables of pressure and entropy. The entropy fluctuations relax within a time scale determined by the thermal conductivity and by the specific heat at constant pressure. In contrast to the fluctuations in the pressure which decay through sound waves, these fluctuations are not propagating. The dynamic susceptibility (Mountain 1966), according to (2.7), allows the determination of the scattering function. The entropy fluctuations cause a central, quasielastically broadened peak and the pressure fluctuations give rise to two side peaks, called the Brillouin doublet.

At shorter wavelengths, the sound waves are strongly damped and the side peaks disappear, leaving only the central peak. The width of this peak varies: it first increases with Q, then it decreases due to "de Gennes narrowing"(de Gennes 1959) near Q_0, which gives the position of the main peak in the static structure factor $S(Q)$. At larger Q the spectrum broadens again. In this region, a liquid is treated as an ideal gas (the so-called particle limit).

In the last years new interest has led to improved and more detailed theoretical descriptions of liquids (Zippelius and Götze 1978; Bosse, Götze and Lücke 1978; Sjögren 1979, 1980; de Schepper 1980, Wang and Overhauser 1988), which were partly stimulated by molecular dynamic calculations (Levesque, Verlet and Kürkijarvi 1973; Rahman 1974; Alley and Alder 1983).

Only a few inelastic neutron experiments have been reported on liquids, such as on sodium (Cocking 1967), on argon (Sköld and Rowe 1974), on rubidium (Copley and Rowe 1974; Suck and Gläser 1975) and on cesium

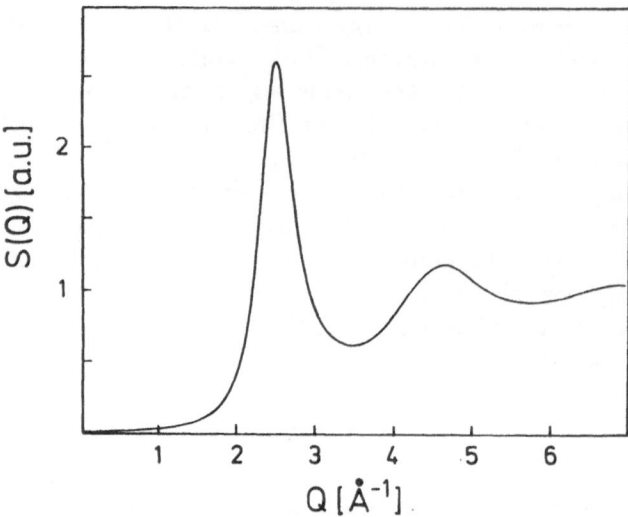

Fig. 6.15. Static structure factor $S(Q)$ for liquid lithium at 600 K as a function of the wave vector drawn according to tabulated values by Waseda (1980)

(Bodensteiner 1990). This situation has motivated the application of inelastic X-ray scattering in studies of liquids.

Before discussing the first results of inelastic X-ray scattering on liquids, one theoretical model will be presented in some detail for later comparison with the experimental data. Following (2.7), the scattering function can be obtained via the general susceptibility $\chi(\boldsymbol{Q},\omega)$. The viscoelastic model derived and discussed by Lovesey (1971), Copley and Lovesey (1975), Hansen and McDonald (1976) and Lovesey (1984) for a monoatomic fluid includes the behavior of the response in the hydrodynamic limit, as discussed before, with the decay of the fluctuations in the particle density through local pressure and entropy variations. Beyond this limit, which means at intermediate wavevectors and frequencies, this model uses generalized equations of motion and considers fluctuations in the particle density and the energy density. Besides a wavevector dependent susceptibility for the particle density, a wavevector dependent specific heat is considered. This theory leads to the scattering function (Lovesey 1984)

$$S(\boldsymbol{Q},\omega) = \frac{1}{\pi} \cdot \frac{\omega\beta}{1 - e^{-\beta\hbar\omega}} \cdot \frac{S(Q)\,\omega_0^2\,(\omega_1^2 - \omega_0^2)\,\tau_r}{[\omega\,\tau_r\,(\omega^2 - \omega_1^2)]^2 + (\omega^2 - \omega_0^2)^2}, \tag{6.6}$$

where $S(Q)$ is the static structure factor and ω_0^2 and ω_1^2 are the normalized second and fourth moments of the particle response function, respectively, and τ_r is a relaxation time. The scattering law (6.6) satisfies the condition of detailed balance and reveals a three peak structure with a central peak and two side peaks, the Brillouin doublet of the hydrodynamic limit, as discussed before. The static structure factor $S(Q)$ is related to the spatial Fourier transform of the pair distribution function. Figure 6.15 shows the static structure factor of lithium as drawn from values tabulated by Waseda (1980).

The main peak in $S(Q)$ gives information on the mean nearest neighbor distance R_0, because its position is at a wavelength $Q_0 \simeq 2\pi/R_0$. According to computer simulation results with hard spheres (Hansen and McDonald 1976), the form of the first peak and its slope to higher Q values is determined by the repulsive part of the pair potential.

The normalized second moment of the particle response function ω_0^2 describes the variation of the energy spread and it depends on the structure factor by

$$\omega_0^2 = \frac{Q^2}{m\beta S(Q)} \cdot \tag{6.7}$$

ω_1^2, the normalized fourth moment of the particle response function, can be derived under the assumption of a single pair potential and the approximation of Einstein oscillators with frequency ω_e,

$$\omega_1^2 = \left(\frac{3Q^2}{m\beta}\right) + \omega_e^2 \cdot \left[1 - \frac{3\sin(QR_0)}{(QR_0)} - \frac{6\cos(QR_0)}{(QR_0)^2} + \frac{6\sin(QR_0)}{(QR_0)^3}\right] \cdot \tag{6.8}$$

For $\beta \to \infty$, at low temperatures, ω_1^2 approaches the eigenfrequency relation of a solid and, generally, for a dense fluid the frequency ω_e is of the order of the maximum frequency in the corresponding solid.

The function τ_r describes a Maxwell relaxation time of the atoms diffusing freely in the fluid. Their velocities are influenced by frictional forces of the fluid, random forces due to collisions with other atoms. Results of computer simulations of the velocity correlations in dense fluids reveal negative correlations that, in addition, have to be considered in the velocity auto-correlation function due to the very highly correlated motion in dense fluids (Lovesey 1984). Thus, in the viscoelastic model the relaxation time τ_r is wavevector dependent and can be derived from the velocity auto-correlation function.

At sufficiently high frequencies the fluid behaves elastically and the velocity of the collective oscillations of the atoms can be described by $[(B + \frac{4}{3}G)/m\rho_0]^{1/2}$, B is the bulk modulus and G is the rigidity modulus. This leads to a velocity which is, in the model, larger than the ordinary sound velocity $(\chi_s m\rho_0)^{-1/2}$, determined by the adiabatic compressibility χ_s.

A useful estimate of the relaxation time τ_r can be obtained from the fact that $S(Q, \omega = 0)$, for the limit of large Q, has to obey the Boltzmann statistics of a single free particle with a Gaussian frequency behavior:

$$\frac{1}{\tau_r} = 2 \cdot \left(\frac{\omega_1^2 - \omega_0^2}{\pi}\right)^{\frac{1}{2}} \cdot \tag{6.9}$$

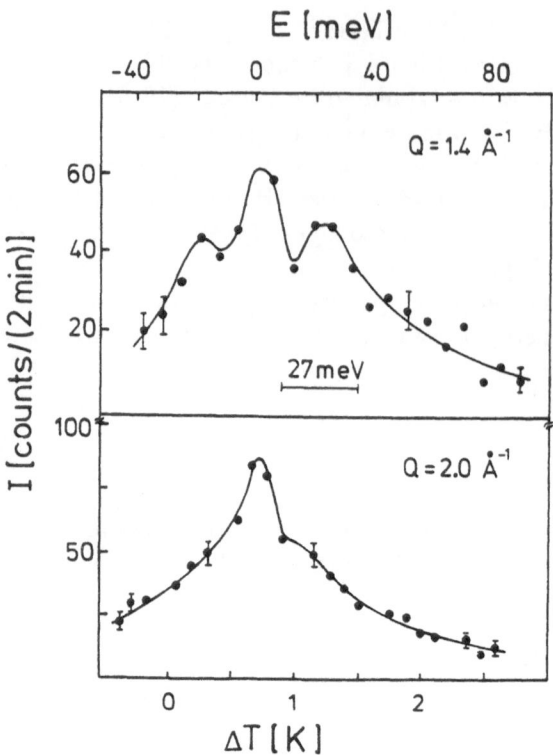

Fig. 6.16. X-ray intensity scattered from liquid lithium at 600 K in constant-Q scans at 1.4 Å^{-1} and 2.0 Å^{-1}. The *lines* are drawn to guide the eye

For the investigations of liquids with inelastic X-ray scattering at INE-LAX liquid lithium was chosen. The experiments were started with a moderate energy resolution $\delta E = 27$ meV (Gauss 1989; Burkel, Gauss, Illini and Peisl 1990). They are at present being continued with an improved value of $\delta E = 12$ meV (Sinn, Burkel 1991). In the performed experiments the liquid state of the lithium sample was always controlled by the measurement of the elastic scattering intensity over a large Q range. Figure 6.16 shows the scattered X-ray intensity at a temperature of 600 K as a function of the temperature difference and the corresponding energy transfer for Q values of 1.4 Å^{-1} and 2.0 Å^{-1}. The improved energy resolution allowed measurements at even smaller Q values such as $Q = 0.4$ Å^{-1} with the intensity distribution shown in Fig. 6.17. Another scan is shown for the wavevector $Q = 3.2$ Å^{-1} in Fig. 6.18.

All spectra clearly demonstrate the expected excitation scheme of a quasi-elastic line with pronounced side peaks or shoulders. The energy transfers for these collective denity modes, drawn in an energy transfer-wavevector diagram in Fig. 6.19 demonstrate the dispersion of these density modes. The adiabatic velocity of sound taken from Ruppersberg and Speicher (1976) is shown as well. The position of the first maximum of the observed dispersion relation corresponds to $Q_0/2$, with Q_0 being the position of the first peak of the static structure factor of liquid lithium. The line drawn in Fig. 6.19

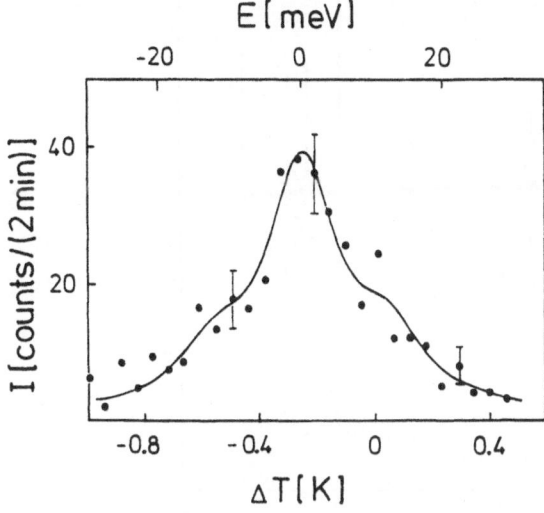

Fig. 6.17. X-ray intensity scattered from liquid lithium in a constant-Q scan at 0.4 Å^{-1}. The *line* represents a fit of three Lorentz curves with widths of 12 meV at half height

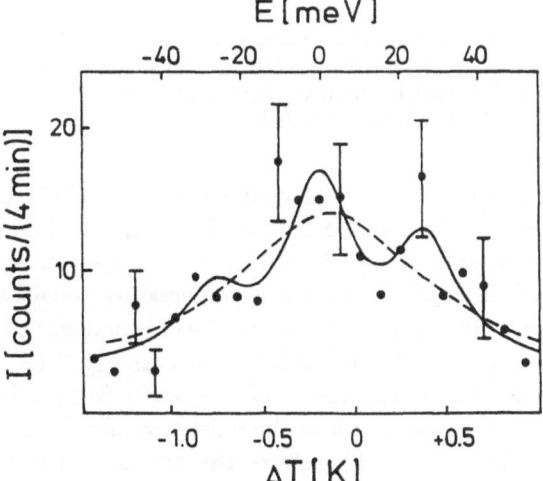

Fig. 6.18. X-ray intensity scattered from liquid lithium in a constant-Q scan at 3.2 Å^{-1}. The *solid line* represents a fit of three Lorentz curves and the *dashed line* is a fit with a single Lorentz curve

was calculated according to (6.6) by deriving the variation of the excitation energy for the sideband peaks (Sinn 1991). A hard sphere radius of $R_0 = 2.72$ Å and a frequency $\omega_e = 22.4$ meV were used for these calculations. There is good agreement between the calculated curve and the X-ray data within the error bars. Even at high Q values, there is remarkable agreement. In addition to the first maximum in $S(Q,\omega)$ there are indications for a second one at higher Q values. This means that the density modes are not overdamped in this Q region. More detailed discussion on the appearance of such propagating modes using a condition given by Lovesey (1971) will be given in Sinn and Burkel (1991).

Another analysis of the data was also attempted. Instead of separating them into three maxima, the experimental scattering distributions were fitted

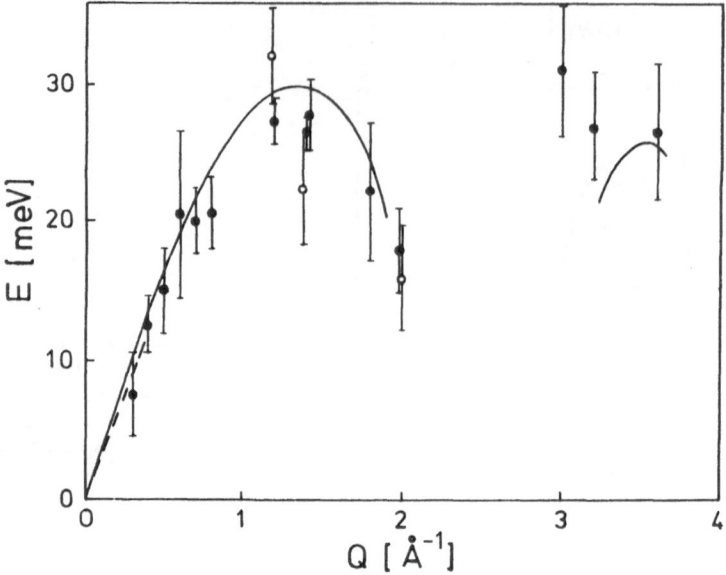

Fig. 6.19. Dispersion of the collective density modes in lithium at 600 K derived from the constant-Q scans in an E-Q diagram. (•) represent data obtained with improved energy resolution. The *curve* is calculated as discussed in the text (Sinn 1991). The *dashed line* represents a slope, according to the adiabatic velocity of sound

with a single Lorentzian curve. An example is shown for $Q = 3.2$ Å^{-1} in Fig. 6.18. The variation of the halfwidth of the Lorentzian fit, determined for a wide Q-range, is given in Fig. 6.20 after deconvolution from the experimental resolution of 12 meV. The halfwidth of the fit curve is certainly sensitive to the positions of inelastic sidebands and to the quasielastic broadening of the scattering intensities. As expected, the behavior shown in Fig. 6.19 is reproduced very well. It supports the interpretation of the scattering profile in Fig. 6.18 assuming three distinct maxima due to the existence of collective propagating density modes in this regime. Therefore, the approach to the ideal gas limit is only expected at still higher Q values.

Nevertheless, an attempt to quantitatively reproduce the data can be made by using the second moment of the particle response function ω_0^2, as defined by (6.7), which describes the variation of the energy spread. The calculated function (Sinn 1991) is drawn in Fig. 6.20. It approaches the gas limit of a free particle at large Q values (*dashed line* in Fig. 6.20). Due to the dependence of ω_0^2 on $S(Q)$ this behaviour is obvious, because $S(Q) \to 1$ for the gas limit. There is very good agreement between the calculated behaviour and the data. At the moment, only additional data collection can help to improve the statistics of the spectra and to reduce the error bars.

From the inelastic X-ray scattering data, obtained so far, it is obvious that in this Q-range the liquid has a solid state like behavior with propagating waves. Around each atom of the liquid a correlation region with a radius of

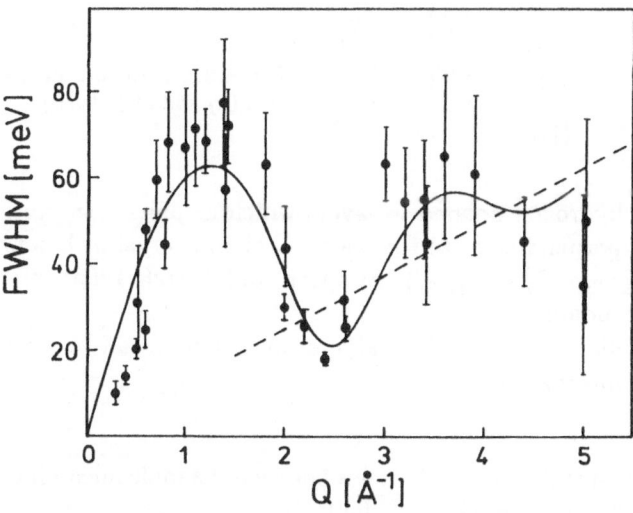

Fig. 6.20. Dependence of the energy halfwidth of an asssumed Lorentzian fit to the X-ray scattering data as a function of the momentum transfer. The data are deconvoluted, assuming an instrumental resolution of 12 meV. The *line* drawn is calculated by Sinn (1991), as discussed in the text. The *dashed line* indicates the frequency behavior in the gas limit

several atomic distances can be expected. In this range collective coupled vibrations similar to phonons might exist.

More sophisticated interpretations based on the discussions of Söderström, Dahlborg and Gudowsky (1985) and Egelstaff and Gläser (1985), and important conclusions might be expected in the future with an improved physical understanding.

6.3 Molecular Vibrations in Amino Acids

It is a particularly challenging idea to apply the technique of inelastic X-ray scattering to even more complex systems than discussed so far. Molecular crystals, in particular, those important in biological systems, were chosen as starting systems. Biological macromolecules are normally too complex to be studied directly. Yet, it is possible to examine their smaller constituent molecules and to extrapolate the collected information on the molecular configuration and on the nature of hydrogen bonding in macromolecules.

The principal molecular constituents of proteins are amino acids. The two simplest of these are glycine and alanine. Their structures are shown in Fig. 6.21.

For the inelastic X-ray studies single crystals of both amino acids were grown by Strohmeier (1989) in collaboration with M. Bartunik at the Max-Planck Institute, Hamburg. Glycine crystallizes in its α form (Marsh 1958), which is monoclinic (space group C_{2h}^5-P2_1/n) with four molecules in the unit

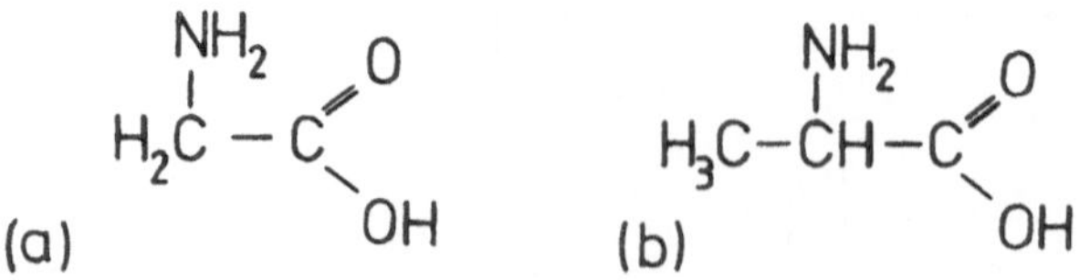

Fig. 6.21. Structure of glycine (a) and alanine (b)

cell. The molecules are hydrogen bonded in layers which as pairs form anti-parallel "double layers" perpendicular to the b axis. L-alanine crystals belong to the rhombic space group D_{2h}^4-($P_{2_1 2_1 2_1}$) (Simpson and Marsh 1966) with almost similar hydrogen bonds.

The modes of molecular crystals such as glycine and alanine can be separated into external and internal modes:

- External modes
  These are vibrational and rotational modes between the molecules which are treated as rigid bodies making up the lattice.

- Internal modes
  These are motions within the molecules, like rotations of parts of the molecule or bending or stretching of molecular bonds.

However, this separation is only possible if intramolecular forces are much stronger than intermolecular ones. Otherwise, mode coupling is present.

The internal modes are of particular interest for inelastic X-ray scattering because of their high frequencies, which makes them difficult to access by neutron scattering. Brillouin scattering or infrared absorption investigations can be performed, but they are restricted to almost zero momentum transfer, as mentioned earlier. Therefore, the determination of the dispersion of such modes – normally treated as dispersionless – by inelastic X-ray scattering could reveal new information.

The investigations on glycine and alanine were performed with an energy resolution of about 17 meV (Strohmeier, Burkel and Peisl 1989). Typical energy scans for alanine are shown in Fig. 6.22 and Fig. 6.23 and for glycine in Fig. 6.24 for energy transfers up to 500 meV. The scattered intensity is always shown as a function of the temperature difference between the analyzer and the monochromator and the corresponding energy transfer. The data points are averaged over at least five scans. All scans show more or less distinct excitations. The energy scan for alanine at $Q = (0\ 12.3\ 0)$ in Fig. 6.22 shows elastically scattered intensity together with distinct energy gain and energy loss signals. Figure 6.23 for $Q = (0\ 12.8\ 0)$ demonstrates the attempt to fit part of the excitation spectrum with Lorentz curves and it clearly supports excitation energies of about 80 and 100 meV. The weak scattering intensity in Fig. 6.24 is an example of more or less distinct excitations. The interpretation of the barely visible maxima can be done at present only by comparing them with results of other kinds of experiments, like Raman and Brillouin scattering, infrared absorption, neutron scattering and with theoretical models.

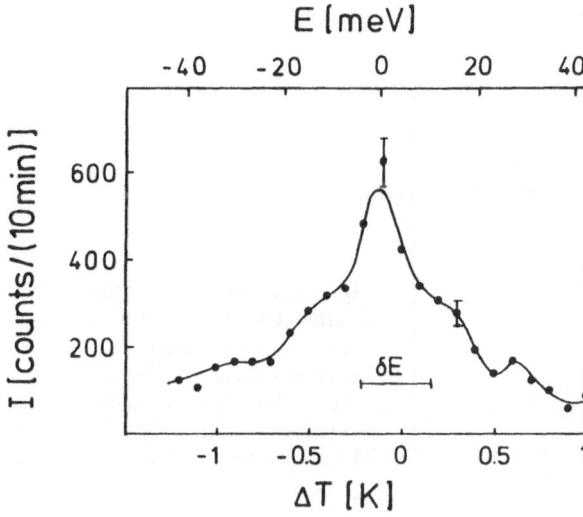

Fig. 6.22. Intensity scattered from alanine at $Q = (0\ 12.3\ 0)$ as a function of the temperature difference and the energy transfer. The *line* is drawn to guide the eye

The scattering profile of glycine in Fig. 6.24 allows an interpretation with excitations at 260, 300, 350, 400 and 460 ± 10 meV. According to results from Raman scattering (Balasuramanian, Krishnan and Iitaka 1962), the mode at 350 meV can be correlated with an N - H $\cdots$ C oscillation, which is an oscillation of the hydrogen atom between a covalent and a hydrogen bond. In this case one would expect the mode to be dispersionless. The mode at 400 meV can be a NH_3 stretching mode, showing an energy of 390 meV in the Raman spectra. This might indicate a minor dispersion of this mode. The maximum at 460 meV has no equivalent in the literature. It is probably due to a higher order excitation.

An attempt to explain the excitations visible around 260 and 300 meV in Fig. 6.24 failed. Raman scattering and infrared absorption indicate that no modes are excited in the energy interval of 230 to 315 meV. One possible explanation could be that the X-ray spectra show optical, Raman inactive modes. A definite interpretation can only be made by model calculations. In any case, further measurements with better statistics are planned.

Figure 6.25 shows all the excitation energies found in alanine in the energy range of 50 to 120 meV as a function of the momentum transfer in the $[0\ \xi\ 0]$ direction for longitudinal modes, as a first attempt to demonstrate the dispersion in this regime. Literature values from neutron scattering (Gupta and Krishnan 1970), Raman and infrared spectroscopies (Wang and Storms 1971) and model calculations (Srivastava and Gupta 1972) are indicated as well. The observed modes marked by I – IV will be discussed in the following:

  – Mode I is described in different ways in the literature. According to model calculations it is a mixture of $(N^+-C_\alpha-CH_3)$ bending and $(C=O)$

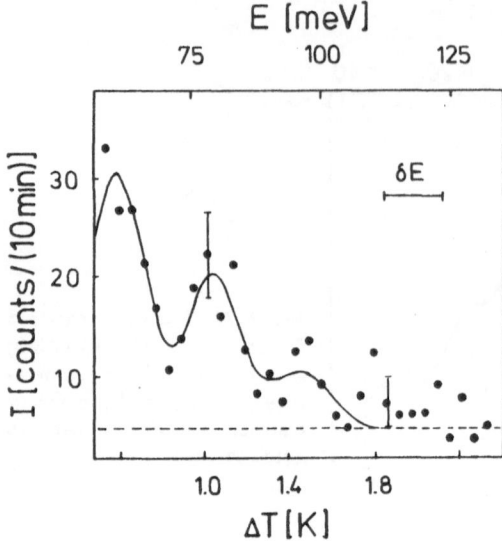

Fig. 6.23. Intensity scattered from alanine at $Q = (0\ 12.8\ 0)$ as a function of the temperature difference and the energy transfer. The *solid line* is a fit of three Lorentz curves with a minimized energy width of 12 meV. The *dashed line* gives the background signal

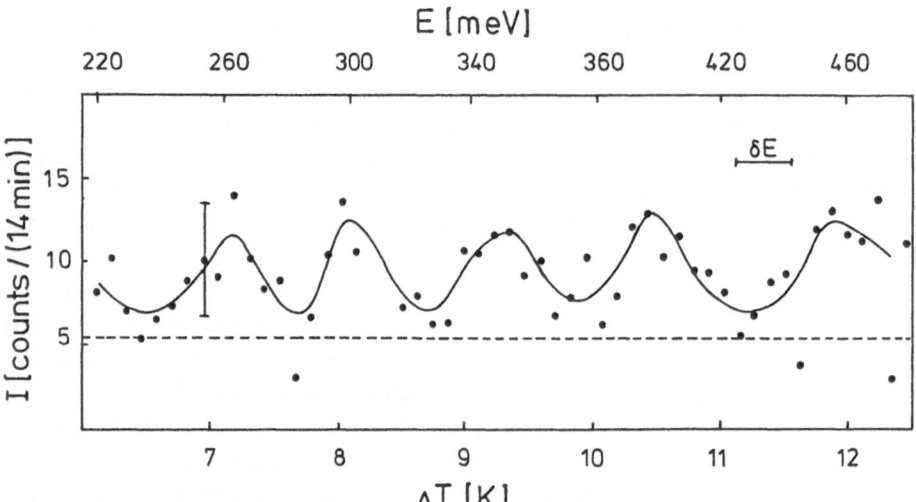

Fig. 6.24. Scattering intensity of glycine at $Q = (0\ 0\ 6.5)$ as a function of the temperature difference in the energy range 220 to 480 meV. The *solid line* is drawn to guide the eye and the *dashed line* shows the background signal

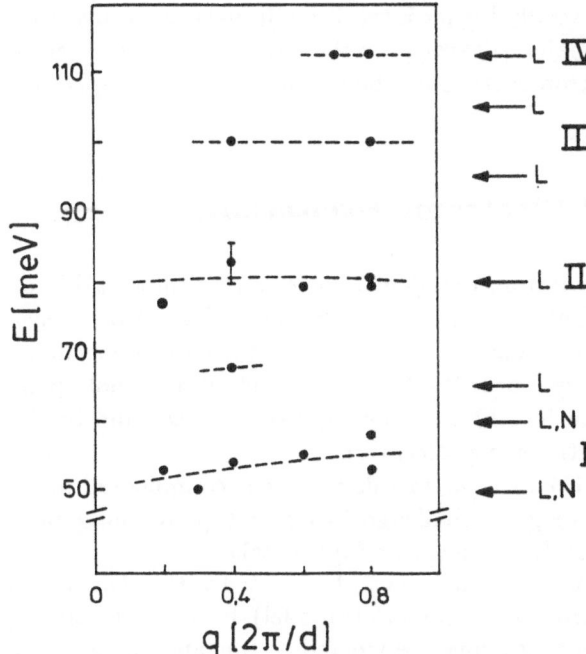

Fig. 6.25. Excitation energies in alanine in an energy-momentum transfer diagram in the energy range of 50 to 120 meV. Literature values for excitation energies are shown as well: (L) Raman and infrared spectroscopy, (N) neutron scattering. Further details on the modes I – IV are discussed in the text

wagging modes and of a torsion of the NH_3^+ group. The calculated energy of 58 meV is assigned by the authors to a mode observed with neutron scattering (Gupta and Krishnan 1970). In earlier works (Fukushima, Onnishi, Shimanouch and Mizushima 1959; Wang and Storms 1971) this mode was discussed as a CCNC deformation.

The energy range of the X-ray data indicates a slight dispersion. This supports the described mixture, with variations in the excitation strength at different momentum transfers.

– Mode II can be identified absolutely as a COO^- wagging motion. There is no dispersion within the error bars.

– Mode III lies between two excitations which are described in the literature. They are classified as COO^- bending and CCNC skeleton stretching (2nd order). The measured energy could result as a mixture of these.

– Mode IV is in excellent agreement with the literature value for an asymmetric streching motion of the CCNC skeleton.

The excitation energy at 68 meV also shown in Fig. 6.25 can be identified as a COO^- wagging motion.

The presented X-ray data are a promising start for the application of the new method to molecular and biological systems. They demonstrate that increased resolution and especially higher intensity will make such investigations very attractive. One major advantage is the capability of using small samples. In contrast to neutron scattering, there is no need for sample deuteration.

6.4 Investigations of Electronic Excitations

Inelastic X-ray scattering is also an excellent tool to obtain information on valence electron dynamics. The dynamic structure factor of an electron system $S(\boldsymbol{Q}, \omega)$ is determined by transitions between occupied and unoccupied one-electron states or, in other words, by the creation of electron-hole pairs induced by a momentum transfer $\hbar \boldsymbol{Q}$ and an energy transfer $\hbar \omega$, and by the correlated motion of the electrons, e.g. plasmons.

In fact, the electrons of even simple metals represent complicated many body systems and it has been a great challenge for a long time to understand the electronic structure of nearly-free electron (sp) metals.

In the simplest approximation electrons can be treated within the model of a non-interacting free electron gas (Sommerfeld model). In this really simple model for a metal, the valence electrons are treated as free and the ions are replaced by a uniform background of positive charge. The electron states are treated as plane-wave functions. An excitation with the energy $\hbar \omega_0$ takes an electron from some occupied state $k < k_F$ within the Fermi sphere to an empty state $k_f = k + Q$ lying outside the sphere (Fig. 6.26a). k_F is the Fermi wavevector. The transition energy is

$$\hbar \omega(\boldsymbol{Q}) = \frac{\hbar^2}{2m} \cdot \left[(\boldsymbol{k} + \boldsymbol{Q})^2 - k^2 \right] = \frac{\hbar^2}{2m} \cdot \left[2 \boldsymbol{k} \boldsymbol{Q} + Q^2 \right] . \tag{6.10}$$

It describes a continuum of excitations and for an excitation energy $E = \hbar \omega_0$ the momentum may change from a value $\hbar k$ to some value between $\hbar Q_{\min}$ and $\hbar Q_{\max}$ (Fig. 6.26a). The spectrum of such allowed intraband excitations is described by

$$0 \qquad < E < \frac{\hbar^2}{2m} \left[2k_F Q + Q^2 \right] \quad \text{if } Q < 2k_F ,$$

$$\frac{\hbar^2}{2m} \left[-2k_F Q + Q^2 \right] < E < \frac{\hbar^2}{2m} \left[2k_F Q + Q^2 \right] \quad \text{if } Q > 2k_F \tag{6.11}$$

and shown in Fig. 6.26c in the energy-wavevector diagram as the region between the two parabolas starting at zero and $2k_F$.

Taking Coulomb interactions of the electrons into account leads to the appearance of collective excitations. It is assumed that the electrons are moving in the average field of the others. The description within such jellium model is done by the self-consistent field method or the random phase approximation

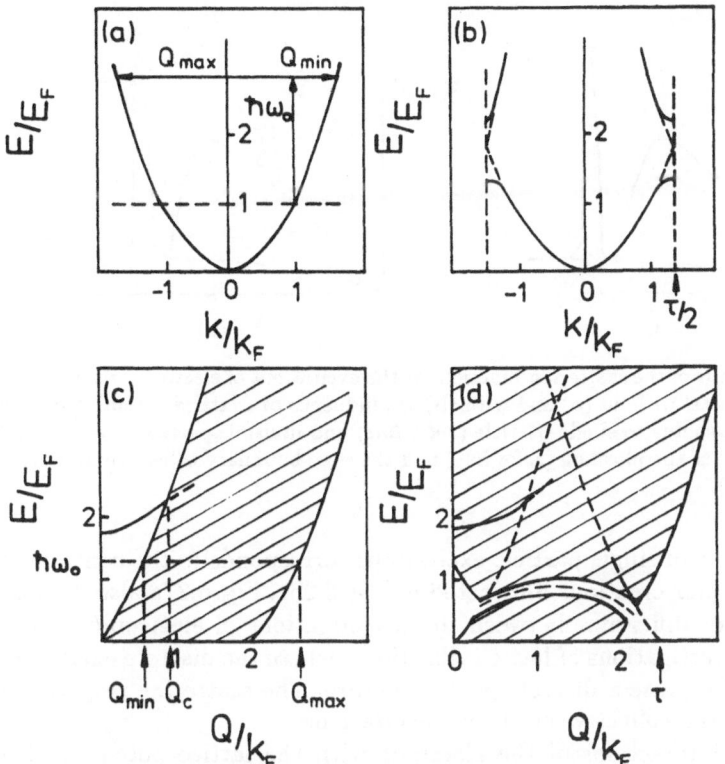

Fig. 6.26. Energies of excitations of a Fermi liquid – without (**a, c**) and with the influence of a lattice potential (**b, d**) – are shown as functions of the wavevector or the momentum transfer, respectively. The *shaded areas* show single-particle excitation regions of intraband transitions (**c**) and intraband and interband transitions (**d**). The *dashed line* located in the gap of the continuum (**d**) represents the zone boundary collective state [after Foo and Hopfield (1968)]. The plasmon is drawn in (**c**) and (**d**) as well

(RPA). The collective excitations are longitudinal plasma vibrations, called plasmons. Their dispersion for small Q within this approximation is shown in Fig. 6.26c. The wavevector Q_c, beyond which the plasmons are not stable and decay via electron-hole excitations, is also indicated.

The behavior of the electrons in such a Fermi liquid is described within the theory of the linear response. Formally, the response of the electrons to a weak external perturbing field is discussed by using the Lindhard dielectric function (Lindhard 1954). As a consequence, it is possible to determine the scattering function, according to (2.8).

A schematic view of the scattering function for the excitations of electrons in the RPA model of a metal is shown in Fig. 6.27a. The dominant contributions are due to the single particle excitations and the collective excitations, i.e. plasmons. A small contribution also arises from multiple particle excitations. In this context, particle means an electron-hole excitation. The

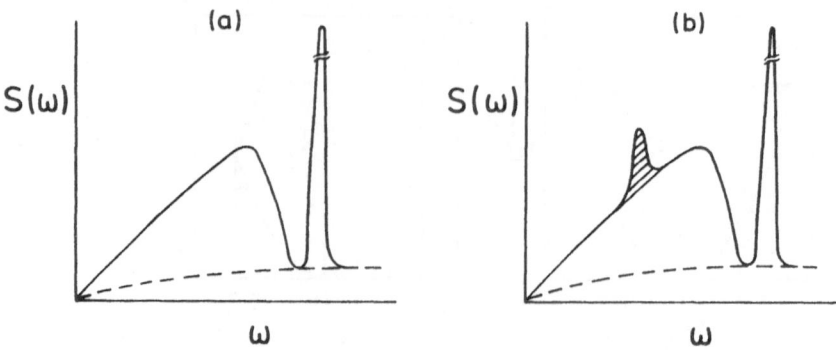

Fig. 6.27. Schematic of the scattering function for the excitations of electrons within in the RPA model of a metal without (a) and with (b) the influence of a lattice potential. Shown are the contributions due to single particle (*solid line*) and multiple particle (*dashed line*) excitations, as well as the plasmon (*solid line*) and the zone boundary collective excitation (*hatched region*, b)

broad distribution of single particle excitations corresponds to the continuum of electron-hole pair creations illustrated in Fig. 6.26c. Fig. 6.27a also demonstrates the major difference between the investigations of electronic excitations and the investigations of lattice vibrations, where the discrete excitation energies normally cause a discrete peak structure. The scattering response is only similar for the collective electronic excitations.

So far, the interactions of the electrons with the lattice potential of a real crystal were neglected in the homogeneous Fermi liquid. However, the influence of the lattice potential with its periodicity leads to the framework of Brillouin zones and to a periodicity in the dielectric function. Band splitting occurs at the zone boundaries. The band structure for a finite sized lattice potential is shown in Fig. 6.26b in a two band model. The excitation spectrum is presented in Fig. 6.26c. Excitations leading from an occupied state to the center of the band gap with $Q \parallel \tau$ are forbidden and determine the dispersion of the excitation gap by

$$\hbar\omega_{\mathrm{Gap}}(Q) = \frac{\hbar^2}{2m}\left[\frac{\tau^2}{2} - \left(\frac{\tau}{2} - Q\right)^2\right] = \frac{\hbar^2}{2m} \cdot Q \cdot (\tau - Q) \, . \tag{6.12}$$

The excitation gap is also indicated in Fig. 6.26c. Besides intraband, interband transitions now become possible in such a nearly free-electron metal

Foo and Hopfield (1968) were the first to discuss this consequence of lattice influence on the electronic excitation spectrum. They calculated the dielectric function within a simple one-reciprocal-lattice vector model, i.e., within a two band model. They stated that the excitation gap due to the influence of the lattice potential causes an additional zero crossing of the dielectric function and, accordingly, a maximum in the loss function. Therefore, the authors expected that the scattering function (2.6) shows a new collective excitation analogous to the plasmons (Fig. 6.27b). Because of the

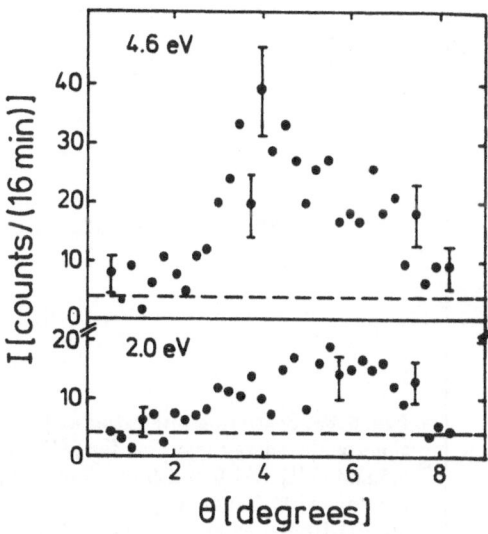

Fig. 6.28. Scattering intensity as a function of the scattering angle for energy transfers of 2.0 and 4.6 eV in lithium. The background level (- - -) is also given

influence of the lattice potential the authors called these excitations zone-boundary collective states (ZBCS). Later, Sturm and Oliveira (1984, 1989) discussed the dispersion of these ZBCS states within the approximation of a two-band model.

The basic plasmon excitations of simple metals were thoroughly investigated by inelastic electron scattering [electron energy loss spectroscopy (EELS)] and by inelastic X-ray scattering. Experimental results on the directional dependence of the plasmon dispersion in Al, Li, Be and other systems (Kloos 1973; Platzman and Eisenberger 1974; Schülke and Lautner 1974; Petri and Otto 1975; Eisenberger, Platzman amd Schmidt 1975; Vradis and Priftis 1985) have shown deviations from the simplifying RPA results. Important observations were the broadening of the plasmon-peaks, changes of their dispersions and the continuation of the peak structures for $Q > Q_c$, the cut-off wavevector of the plasmon, and the appearance of an unexpected second peak. These experimental facts were seen as indications of a strongly correlated electron liquid and of non-negligible influences of the crystal potential (for overviews see Raether 1980 and Schülke et al. 1986).

In addition, inelastic scattering of X-rays (Platzman and Eisenberger 1974; Schülke and Lautner 1974) and of electrons (Zacharias 1975; Batson and Silcox 1983) have shown additional peaks for $Q > Q_c$ at lower excitation energies than the plasmon energies. These peaks were interpreted as zone boundary collective states (Sturm and Oliveira 1984).

The ZBCS states could be observed in inelastic X-ray scattering experiments on single crystalline Li with an energy resolution of about 1 eV (Schülke, Nagasawa and Mourikis 1984; Schülke, Nagasawa, Mourikis and Lanzki 1986), indicating strong anisotropy of these excitations in the (Q, ω) regime. More detailed investigations were performed for single crystalline Al with electron-loss spectroscopy (Petri and Otto 1975).

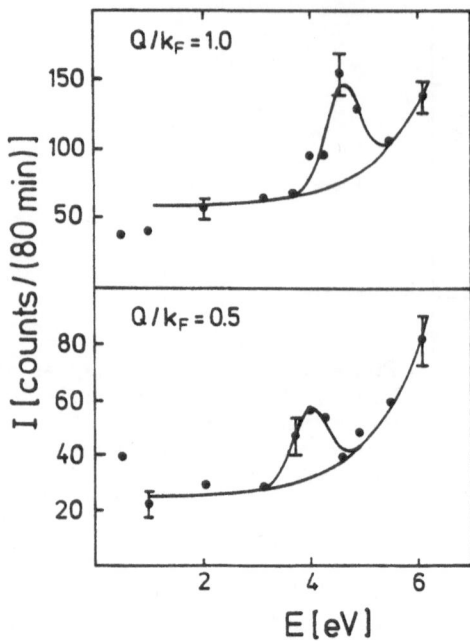

Fig. 6.29. Scattering intensities as functions of the energy transfers for $Q/k_F = 0.5$ and $Q/k_F = 1.0$ in lithium. The *lines* are fits with Gaussian profiles for the observed maxima and exponential functions for the remaining signals in an empirical description (Illini 1991)

The ZBCS excitations in a number of materials are in the energy interval which is accessible with the INELAX instrument. One purpose of our experiments was to investigate these excitations with an energy resolution which was better than used by Schülke, Nagasawa, Mourikis and Lanzki (1986) and to look for indications of fine structure. Because of its small absorption cross section the investigations were performed on lithium (Illini, Burkel and Peisl 1989; Illini 1991). The single crystals were grown with a modified Bridgeman method (Daniels 1960; Illini 1991).

The investigations were done with a photon energy $E_i = 15.8$ keV and with an energy resolution $\delta E = 40$ meV. Due to the high excitation energies, high temperature differences between the monochromator and the analyzer are needed. This requires long time intervals for stabilization (Sect. 4.2). Therefore, constant-energy (temperature) scans were performed in the range of $E = 1.0$ to 6.1 eV with steps varying between 0.3 eV and 1.1 eV. Figure 6.28 shows the results, always as sums of three separate scans, at the energy transfers of 2.0 eV and at 4.6 eV. There are strong variations of the scattering intensities as function of the scattering angles and consequently of the momentum transfers, especially at 4.6 eV. The scattering signals, with 2 to 3 counts per minute, are drastically weaker than the phonon scattering intensities discussed in Sect. 6.1. They are, nevertheless, well separated from the background signal, as indicated in Fig. 6.28.

After normalization, the scattering intensities from all scans can also be shown in the familiar way as functions of the energy transfers. Figure 6.29

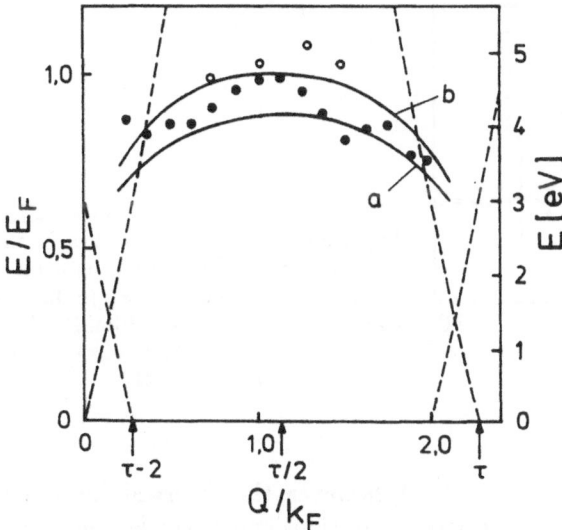

Fig. 6.30. Dispersion curve of the experimentally observed collective electronic excitations in lithium in the [1 1 0] direction. The *broken lines* give the range of intraband transitions of a free-electron gas in a weak lattice potential. (•) INELAX results, (o) experimental results of Schülke, Nagasawa, Mourikis and Lanzki (1986). For comparison, the calculated dispersion curves of the ZBCS states are shown, according to Sturm and Oliveira (1989) with the parameters taken (a) from Ching and Callaway (1974) and (b) from Bross and Bohn (1975)

gives the results for $Q/k_F = 0.50$ and for $Q/k_F = 1.0$. Distinct maxima of the scattered intensity are visible on top of an increasing scattering signal. The drawn lines are fits with Gaussian profiles for the maxima and exponential functions for the remaining signals in an empirical description (according to Illini 1991). The maxima can be correlated with the discussed ZBCS excitations. The intensity increase at higher energies can be due to interband transitions or collective excitations.

Figure 6.30 shows the observed excitation energies in an energy-momentum transfer diagram and demonstrates the dispersion of the collective electronic excitation in the [1 1 0] direction. There is a weak overall dispersion with an obvious fine structure of distinct oscillations, a phenomenon which is observed for the first time. There is symmetry to $\tau/2$, the position of the zone boundary, for both effects. The drawn lines are theoretical dispersion values for the ZBCS states, derived by Sturm and Oliveira (1989) from bandstructure calculations of Ching and Callaway (1974, *solid line* (a) in Fig. 6.30) and Bross and Bohn (1975, *solid line* (b) in Fig. 6.30) under the assumption of a simple two-band model.

The INELAX results show slightly lower excitation energies than the data found by Schülke, Nagasawa, Mourikis and Lanzki (1986). If different sets of parameters are used for calculating the theoretical values, the observable

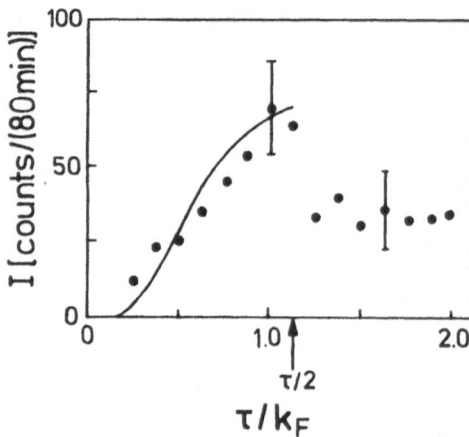

Fig. 6.31. Scattering intensities of the collective electron excitation as a function of the reduced wavevector in the first and second Brillouin zone (Illini 1991) together with ZBCS calculations (*line*) by Sturm and Oliveira (1984)

discrepancies might diminish (Sturm 1990); however, the observed fine structure of the dispersion of collective electron excitations in lithium demonstrates strong influences from the crystal potential. For the theoretical approach it might become necessary to give up the simplifying picture of ZBCS and to go beyond the random phase approximation and to perform more sophisticated, complete calculations (Bross 1978; Seoud 1983) of $\varepsilon(\mathbf{Q}, \omega)$ taking into account the total crystal potential and short range correlations of the electrons.

The intensities of the observed collective electronic excitations are shown in Fig. 6.31 together with theoretical predictions of Sturm and Oliveira (1989) for the first Brillouin zone. The experimental data are corrected for the Q dependence of the scattering volume and for the Q^2 dependence (2.8) of the scattering function. The agreement is again not satisfying. Improvements are necessary and should also be able to explain the different intensity behaviour in the second Brillouin zone. The intensity behavior very much resembles the behavior of the structure factor of a quantum liquid like ^{4}He (Achter and Meyer 1969). This might stimulate further studies of the Fermi liquids and of the strong influence of the the lattice potential. Despite this influence it might reveal liquid-like behavior in its static and dynamical density correlations (see also Platzman 1974).

In future, the increased resolution in inelastic X-ray scattering will provide an excellent tool to study the fine structure of the collective electron excitation states. The study of the energy widths of the excitations will reveal information on damping and the decay mechanism. At present, the observed intrinsic energy width of the excitations is about 1.0 eV (Fig. 6.29), well above the experimental energy resolution of 38 meV, but the statistics are not good enough to detect a possible Q dependence.

Such inelastic X-ray scattering experiments will deliver information complementary to studies with electron scattering. When comparing the methods it is important to discuss the double differential scattering cross section for both of them. For X-ray scattering it is given by (2.8) and (2.9). For electron scattering (2.12) together with the energy-loss function gives

$$\frac{d^2\sigma}{d\Omega\, d\omega_{\rm f}} \propto \frac{1}{Q^2} \cdot {\rm Im}\left[\frac{-1}{\varepsilon\,(Q,\omega)}\right] \ . \tag{6.13}$$

(6.13) demonstrates that electron scattering is a useful experimental tool only for small momentum transfers $\hbar Q$, because of the dependence on Q^{-2}. At higher Q values the scattering intensity is strongly decreasing, and, in addition, various multiple scattering processes are excited, thus making the interpretation of data tedious. In contrast, due to the dependence on Q^2 [see (2.8)], the X-ray scattering intensity is strongly increasing for higher Q values. Therefore, inelastic X-ray scattering is not only complementary to electron energy loss spectroscopy, but opens new areas of application.

It is becoming increasingly interesting to test the knowledge gained so far on electron systems which are more complex than those of, for instance, simple d-band metals. In such complex systems the nearly-free electron gas approximation fails since the electrons are more localized, as the f-electrons in intermediate valence systems, for example. In these cases the understanding of the dielectric properties is not complete and, therefore, investigations of the dynamic response function of such systems with inelastic X-ray scattering will certainly yield valuable information. However, a large penetration depth of the X-rays will be necessary for these studies.

The creation of electron-hole pairs across the band gap in semiconductors, especially in semiconductors with an indirect band gap like silicon, would also be an interesting application of X-ray investigations. Important information for band structure engineering could be gained.

7. The Future of the Technique

7.1 Further Applications

7.1.1 Elastic Scattering with High Energy Resolution

In all the applications of inelastic X-ray scattering the major focus has been on the analysis of the energy resolved inelastic scattering spectra. But it is also possible to study the elastically scattered intensity and to separate the disturbing inelastic background, consisting mainly of thermal diffuse scattering. Elastic scattering intensity contains information on the strength and symmetry of displacements caused by defects in crystals. This scattering intensity is called Huang diffuse scattering (HDS) in the vicinity of Bragg points (Peisl, Spalt and Waidelich 1967; Trinkaus 1971; Dederichs 1973) and Zwischenreflexstreuung (Haubold 1975) elsewhere. This method of analyzing the intensity is highly developed. However, if defects modify the dynamics of the host lattice drastically or if investigations have to be performed at higher temperatures, experimental separation of thermal diffuse intensity is necessary. Until now this has required switching from X-ray to neutron scattering (Burkel et al 1979; Burkel 1982).

Studying the diffuse quasielastic intensity (Burkel et al 1988, Wochner et al 1990) of fast relaxing defect states is an interesting domain, which can also be performed time resolved, by using the time structure of the synchrotron beam.

7.1.2 Inelastic Scattering Under High Pressure

A completely different area of study is the behavior of condensed matter under pressure. In this field X-ray scattering has a clear advantage over neutron scattering. The reason is the need for high-performance pressure cells, which can be easily manufactured only for small samples.

The study of the phonon dispersions of solids under pressure will reveal information on variations of the Grüneisenparameter. A further area will be the study of liquids and the change of their collective excitations up to the pressure of solidification.

7.1.3 Anomalous Inelastic X-ray Scattering

An important new possibility in X-ray scattering created by synchrotron radiation focuses on anomalous scattering. The capability to tune the primary energy of the photons in a scattering experiment allows large variations of the atomic scattering amplitudes (up to 30%) for X-rays close to the absorption edges for the particular atoms.

The atomic form factor can be separated into contributions dependent or independent of the primary photon energy E_i:

$$f(Q, E_i) = f(Q) + f'(E_i) + i f''(E_i) , \tag{7.1}$$

with $f(Q)$ being the atomic form factor far above the K-absorption edge and used so far in this text. $f'(E_i)$ and $f''(E_i)$ are the contributions due to excitations of electrons in the inner shells. If E_i is in the energy range of the atomic resonance frequencies, the photons can create real and virtual excitations of the core electrons. Therefore, photons will be absorbed and emitted again.

This energy dependence of the atomic form factor on the photon energy allows one to vary the contrast in scattering experiments and to enhance scattering contributions of a certain type of atoms (Stuhrmann 1980) and even to perform resonance scattering experiments. In inelastic X-ray scattering this means, for instance, that certain phonon modes or particular molecular vibrational modes are enhanced. In such resonance studies, however, it might be necessary to include an electron-phonon coupling term and to discuss the scattering in higher order perturbation theory.

It is obvious that, at the moment, the inelastic X-ray experiments with INELAX are restricted to discrete levels of the primary energy E_i. However, the possibilities of varying the temperature and the types of Bragg reflections at the monochromator and analyzer will certainly lead to interesting applications in the near future.

7.1.4 Inelastic Scattering with Polarization Analysis

Throughout this review the influence of the polarization of the photon beam on the scattering process was ignored. Only the Thomson term, which expresses the scattering at the charge density, was discussed. However, in the classical view, the electron is accelerated by the incident field and gives rise to electric and magnetic dipole reradiation. Taking both contributions into account, the double differential scattering cross section (2.4) can be written as (Platzman and Tzoar 1970):

$$\frac{d\sigma}{d\Omega d\sigma_f} = \left[\frac{e^2}{mc^2} \right]^2 \cdot S(\boldsymbol{Q}, \omega) , \tag{7.2}$$

with the expanded scattering function $S(\boldsymbol{Q}, \omega)$ (2.5)

$$S(\boldsymbol{Q}, \omega) = \sum_{\varUpsilon_{i,f}} \left| \left\langle \varUpsilon_f \right| \sum_j M_j e^{i\boldsymbol{Q}\boldsymbol{r}_j} \left| \varUpsilon_i \right\rangle \right|^2 \delta(E_{\varUpsilon_f} - E_{\varUpsilon_i} + \omega) \, . \tag{7.3}$$

M_j is the amplitude for the scattering from a single electron with position $\boldsymbol{r}_j$ and spin $\boldsymbol{\sigma}_j$. In a simplified approach taking only leading terms, the amplitude M_j is written as

$$M_j = A + iC\boldsymbol{\sigma}_j \, , \tag{7.4}$$

with

$$A = \boldsymbol{e}_i \cdot \boldsymbol{e}_f \, , \tag{7.5}$$

and

$$\begin{aligned}
C &= -\frac{\hbar\omega_i}{mc^2} \left[\boldsymbol{e}_i \cdot \boldsymbol{e}_f \left(\hat{k}_i \times \hat{k}_f \right) - \frac{1}{2} \left(\hat{Q}\hat{Q} \right) (\boldsymbol{e}_i \times \boldsymbol{e}_f) \right. \\
&\quad \left. - \hat{Q} \times \left(\hat{Q} \times \boldsymbol{e}_i \times \boldsymbol{e}_f \right) \right].
\end{aligned} \tag{7.6}$$

$\hat{k}_i, \hat{k}_f$ are normalized wave vectors. The leading part in (7.4) describes the scattering from the electronic charge density, already discussed as the Thomson scattering term, and the second part represents the scattering from the atomic spin magnetization densities. For a detailed description of the cross section including also the orbital magnetization density, it is referred to Blume (1985). The cross section from (7.2), correspondingly, shows three contributions, which are pure charge scattering, pure magnetic scattering and the interference between both.

The magnitude of the interference term is reduced by the prefactor of C in (7.6) and its size is about $\hbar\omega_i/mc^2 \approx 10^{-2}$. In analogy, the pure magnetic term is reduced by a factor of 10^{-4}. Nevertheless, these small spin dependent parts can be observed experimentally since the magnitude of the Thomson contribution can be varied and even extinguished by appropriate selection of the polarization vectors. By using circularly polarized photons, for instance from an asymmetric wiggler, C is complex and gives real contributions to the double differential scattering cross section. It opens the possibility to study the spin density and the magnetic response functions of different systems.

The understanding of magnetic phenomena requires the knowledge of the whole magnetic excitation spectrum. Magnetism in condensed matter is produced by unpaired electrons. Strong exchange interactions lead to collective excitations, magnons, which merge after a strong dispersion into a broad particle-hole continuum. The excitation energies are as high as several eV. An advantage of the studies of magnetic excitations with X-rays is in the capability to distinguish the orbital- and spin-angular-momentum contributions by polarization analysis (Blume 1985). Such an observation is not possible with neutron scattering.

Due to the small size of the magnetic scattering term the scattering cross section has, including form factors, a ratio of about $4 \cdot 10^{-6}$ between charge and magnetic scattering for X-rays in the keV range. Despite the discussed polarization effects, this means that for the high resolution measurement of contributions due to magnetic scattering more intense sources then presently available are necessary.

But another type of study might already be achievable. The interference between the charge and the magnetic scattering can be used to detect excitations that have no significant charge scattering response, but which are correlated with a certain variation of polarizability. These experiments can be seen in analogy to Brillouin and Raman light scattering.

7.1.5 Inelastic Scattering Under Grazing Incidence

Scattering experiments under conditions of grazing incidence allow one to control the penetration depth in the sample by variation of the glancing angle. This thus provides depth dependent data. Presently, this method is being highly developed for elastic investigations of near surface layers or layered systems. A dramatic increase of the photon flux will also make inelastic scattering experiments under conditions of grazing incidence more feasible.

7.2 The Next Generation of INELAX

Some of the ambitious applications of inelastic X-ray scattering with very high energy resolution will become possible with the higher intensities of future sources combined with the further improvements. For more general applications a larger penetration depth is desirable and, therefore, the energy E_i of the instrument should be in the energy range of 20 to 30 keV. This also provides an improved energy resolution of less than $\delta E = 1$ meV, according to Fig. 2.5. However, in this case the energy resolution is limited by geometrical conditions. It will be useful to switch from the present use of a wiggler source to an undulator source. This will provide a smaller illuminated area on the monochromator, thus limiting the geometrical influence on the energy resolution. In addition, there is no longer a need for a large area focusing device at the monochromator. This will reduce the present intensity loss at this component. For most applications the momentum resolution can be improved with a decreased analyzer area as well.

The ideal source for such an improved instrument will be an asymmetric undulator at a high energy ring, because of the proposed energy range. Figure 7.1 (thanks to P. Gürtler, HASYLAB, Hamburg, for the calculations and for supplying this figure) shows the brightness of the photon beam as a

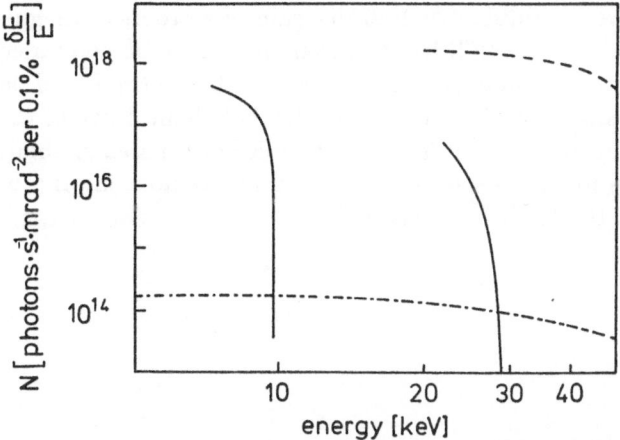

Fig. 7.1. The brightness as a function of energy for the HARWI wiggler at DORIS, Hamburg ($\cdot - \cdot$ *line*), for an undulator at ESRF, Grenoble (fundamental and 3rd harmonic, *solid line*) and for an undulator at PETRA, Hamburg (fundamental, *dashed line*)

function of the energy. It compares the currently used HARWI wiggler at DORIS, Hamburg, with a projected symmetric undulator device at the ESRF, Grenoble, and with a study of an undulator device at PETRA, Hamburg. The parameters which were used for the calculations of the diagrams of the brightness in Fig. 7.1 are shown in Table 7.1. The brightness of a typical ESRF undulator is shown for the fundamental and the third harmonic (ESRF Red Book 1987). For the PETRA undulator only the fundamental is shown. The undulator at PETRA was designed using the parameters of a HASYLAB undulator and only changing the optics from an emittance of 92 nm·rad to 10 nm·rad, ignoring more rigorous improvements to about 1 nm·rad.

Table 7.1. The parameters of the insertion devices at DORIS and PETRA in Hamburg and at ESRF in Grenoble, used for the calculation of the brightness shown in Fig. 7.1

		Wiggler W2 DORIS	Undulator ESRF	Undulator PETRA
Nominal beam energy	GeV	5.3	6	14
Beam current	mA	40	100	60
Horizontal emittance	m·rad		$7 \cdot 10^{-9}$	$10 \cdot 10^{-9}$
Horizontal angular distribution	mrad	0.61	0.015	0.02
Vertical angular distribuion	mrad	0.054	0.008	0.0063
Length	m	2.40	3.4	10
Gap	mm	42	20	11
Period	mm	240	35	33.5

The comparison shows an impressive intensity gain of 4 orders of magnitude in favor of an undulator at PETRA. However, it cannot be concealed that this enormous brightness also causes an increase in the thermal load on the crystals. It is comparable with the present heat load of about 3 kW at the HARWI wiggler, but at the undulator it has to be handled on an area of about 5 mm^2, if a distance of 90 meters is selected. Therefore, a premonochromator will still be necessary for the INELAX instrument.

8. Final Remarks

The INELAX instrument with an energy resolution in the range of meV brought about a breakthrough in X-ray diffraction and spectroscopy experiments with the direct measurement of small energy shifts of photons. It is now possible to resolve phonon scattering with X-rays as well.

This step opened a wide area of research activities in high energy resolution work. Convincing results in single crystal phonon scattering were obtained, demonstrating the power of the method. Experiments with polycrystalline and liquid matter show the competitive power of X-rays compared to neutrons. This is underlined by the detection of molecular vibrations in small biological samples. Investigations of electronic excitations with an increased resolution reveal fine structure, demanding modifications to the theoretical approaches in this field.

Certainly, not all spectra shown here have optimal statistics, but they must be regarded as pilot experiments. The development of the capabilities of high resolution X-ray scattering can be best demonstrated by comparing an inelastic scattering signal taken early in INELAX' life with one from today. Figure 8.1a shows the scattered elastic and inelastic intensity of pyrolythic graphite as a function of the temperature difference in one of the single scans that are added up in Fig. 6.14a. The comparison with a longitudinal phonon in beryllium shown in Fig. 8.1b, measured also in a single scan, clearly shows the dramatic improvements that have been taken place in the last few years.

The attractiveness of inelastic X-ray scattering due to its high resolution and the ease of handling will be increased by the possibility of higher photon fluxes at the sample. The next generation of instruments at new sources like ESRF in Grenoble, or undulators at high energy rings like PEP in Stanford, or eventually at PETRA in Hamburg, will certainly make experiments with an meV energy resolution a standard method in X-ray diffraction and thus increase the number of applications.

In addition, the new sources for synchrotron radiation will also stimulate the development of inelastic X-ray scattering in the ultra high energy resolution range of neV and μeV. The scattering experiments will be performed with the monochromator and analyzer working on the basis of resonant nuclear Bragg scattering with energy widths of 10^{-6} to 10^{-8} eV and the expected development of focusing techniques will supply the sufficient photon flux. This coming generation of inelastic scattering experiments will even allow to use X-rays for the investigations of such dynamics in condensed matter which are now only accessible by its quasielastic scattering effects.

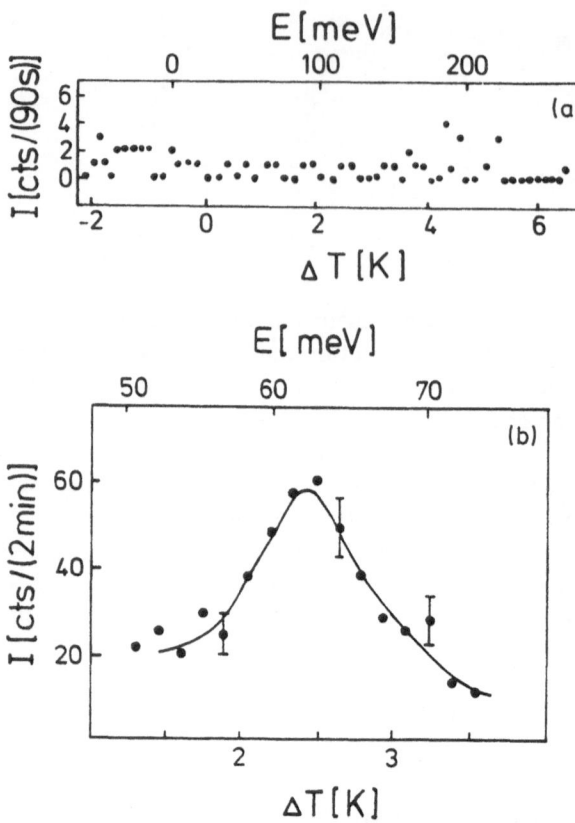

Fig. 8.1. Scattering intensity of an early INELAX scan on pyrolytic graphite as a function of the temperature difference between the analyser and the monochromator, showing elastic and inelastic contributions (a). Scattering intensity of a longitudinal phonon at Q = (0 0 4.5) in beryllium, measured recently, after many instrumentational improvements (b)

Therefore, taking all activities and developments into account energy resolved inelastic X-ray scattering is about to become a convincing powerful method.

List of Tables

References

Achter F. K. and Meyer L., 1969: Phys. Rev. **188**, 291.

Alefeld B., 1966: Sitzungsberichte der Bay. Akademie der Wissenschaften, **109** (Conference proceedings).

Alefeld B., Birr M. and Heidemann A., 1968: in "Neutron Inelastic Scattering." (IAEA Vienna), 381.

Al-Jishi R. and Dresselhaus G., 1982: Phys. Rev. B **26**, 4514.

Alley W. E., Alder B. J., Yip S., 1983: Phys. Rev. **A27**, 3174.

D'Ans-Lax, 1967: Taschenbuch für Chemiker und Physiker, Bd. **1**, 1 (Springer, Berlin-Heidelberg-New York).

van Arkel A. E., 1927: Z. Krist., **27**, 25

Bacon G. E., 1962: "Neutron Diffraction." 2nd ed. (Clarendon Press, Oxford).

Balasubramanian R. S., Krishnan R. S. and Iitaka Y., 1962: Bull. Chem. Soc. Jpn., **35**, 1303.

Batson P. E. and Silcox P. E., 1983: Phys. Rev. **B27**, 5224.

Becker P., Seyfried P. and Siegert H., 1982: Z. Phys B **48**, 17.

Benda Th., Dorner B. and Peisl J., 1983: "Inelastic X-ray Scattering with Very High Energy Resolution." Workshop ESRF. Ed. by Buras B., (CERN).

Bilderback D. H., 1986: Nucl. Instr. and Meth. **A246**, 434.

Bilderback D. H., Lairson B. M., Barbee T. W. Jr., Ice G. E. and Sparks C. J., 1989: Nucl. Instr. and Meth. **208**, 251.

Blume M., 1985: J. Appl. Phys. **57**(1), 3615.

Blumenroeder S., Tirngiebl E., Thompson J. D., Killough P., Smith J. D. and Fisk Z., 1987: Phys. Rev. B **35**, 8840.

Böni J., Axe J. D., Shirane G., Birgenau J., Gabbe D. R., Jenssen H. P., Kastner M. A., Peters C. J., Picone P. J. and Thurston T. R., 1988: Phys. Rev. B **38**, 185.

Bodensteiner T., 1990: Thesis, Technische Universität München.

Bonse U., 1979: "X-Ray Sources." DESY Report No. SR-79/29, Hamburg.

Bosse J., Götze W., Lücke M., 1978: Phys. Rev. A **18**, 1176.

Bottom V. E., 1965: An. Acad. Brasiliera Ciencias **37**, 407.

102   References

Bottom V. E. and Carvalho R. A., 1971: Rev. Sci. Instrum. **42**, 196.

Born M., 1942: Repts. Progr. Phys. **9**, 356.

Bragg W. L., 1913: Proc. Cambridge Phil. Soc. **17**, 43.

Brockhouse B. N., Arase T., Caglioti G., Rao K. R., Woods A. D. B., 1962: Phys. Rev. **128**, 1099.

Bross H. and Bohn G., 1975: Z. Phys. B **20**, 261.

Bross H., 1978: J. Phys. F. **8**, 2631

Brümmer O., Höche H. R. and Nierer J., 1979: Phys. Stat. Sol.(a) **53**, 565.

Brun T., Grimsditch M., Gray K. E., Bhadra R., Maroni V. and Loong C. K., 1987: Phys. Rev. B **35**, 8837.

Burkel E., Guerard B. v., Metzger H., Peisl J. and Zeyen C. M. E., 1979: Z. Phys. B **35**, 227.

Burkel E., 1982: Thesis, University of Munich.

Burkel E. and Peisl J., 1985: unpublished calculations.

Burkel E., Peisl J., Dorner B., 1987: Europhys. Lett. **3**, 957.

Burkel E., Dosch H., Peisl J., Wochner P. and Zeyen C. M. E., 1988: Verhandlungen der DPG **(VI)23**, IV, DY 14.44 (Conference proceedings of the German Physical Society).

Burkel E., Dosch H., Peisl J., Wochner P., Zeyen C. M. E., 1988: Verhandlungen der DPG **(VI)23**, IV, DY 14.44 (Conference proceedings of the German Physical Society).

Burkel E., 1989: "X-Ray Instrumentation in Medicine and Biology, Plasma Physics, Astrophysics and Synchrotron Radiation." Ed. by Rene Benattar, Proc. SPIE Vol. **1140**, 426.

Burkel E., Dorner B., Illini Th. and Peisl J., 1989: Rev. Sci. Instrum. **(60)7**, 1671.

Burkel E., Gaus S., Illini Th. and Peisl J., 1990: PHONONS 89 (Conference proceedings), Vol. **2**, 1436 (World-Scientific, Singapore).

Burkel E., Dorner B., Illini Th. and Peisl J., 1990: J. Appl. Cryst., in press.

Ching W. Y. and Callaway J., 1974: Phys. Rev. B **9**, 5115.

Cocking S. J., 1967: Adv. Phys. **16**, 189.

Cochran W. and Cowley R. A., 1967: *Phonons in Perfect Crystals*, in "Handbuch der Physik." Vol XXV/2a. Ed. by Genzel L. (Springer, Berlin, Heidelberg, New York).

Cooper M. J. and Nathans R., 1967: Acta. Cryst. **23**, 357.

Copley J. R. D., Rowe J. M., 1974: Phys. Rev. A **9**, 1656.

Copley J. R. D. and Lovesey S. W., 1975: Rep. Prog. Phys. **38**, 461.

Curien. H, 1952: Bull. soc. franc. minéral. **75**, 197; Acta Cryst. **5**, 392.

Daniels W., 1960: Phys. Rev. B **119**, 1246.

Debye P., 1913: Verhandlungen der DPG **15**, 678, 738, 857 (*Conference Proceedings of the German Physical Society*).

Debye P., 1914: Ann. der Physik **43**, 49.

Dederichs P. H., 1973: J. Phys. F **3**, 471.

Dix W. R., Glüer C. C., Graeff W., Höhne K. H. and Kupper W., 1982: DESY Report No. SR-82-24.

Dorner B., 1966: Kernforschungsanlage Jülich, Report No. Jül-412-NP.

Dorner B., 1972: Acta. Cryst. A **28**, 319.

Dorner B. and Comes R., 1977: in "Dynamics of Solids and Liquids by Neutron Scattering." Topics in Current Physics, Vol **3**. Ed. by Lovesey S.W. and Springer T. (Springer Berlin), 134.

Dorner B. and Peisl J., 1983: Nucl. Instr. and Meth. **208**, 587.

Dorner B., 1984: "Workshop on High Energy Excitations in Condensed Matter." Los Alamos.

Dorner B., Benda Th., Burkel E. and Peisl J., 1985: "Festkörperprobleme", Advances in Solid State Physics, XXV, (Vieweg Braunschweig) 685.

Dorner B., Burkel E. and Peisl J., 1986: Nucl. Instr. and Meth. in Phys. Res. A **246**, 450.

Dorner B., Burkel E., Illini Th. and Peisl J., 1987: Z. Physik B Condensed Matter **69**, 179.

Dorner B., Burkel E., Illini Th. and Peisl J., 1990: PHONONS 89 (Conference proceedings), Vol.**2**, 1405 (World-Scientific, Singapore).

Eisenberger P., Platzman P. M. and Schmidt P., 1975: Phys. Rev. Lett. **34**, 18

Egelstaff P. A. and Gläser W., 1985: Phys. Rev. A **31**, 3802.

Egger H., Hofmann W. and Kalus J., 1984: Appl. Phys. A **35**, 41-45.

ESRF, Red Book, 1987: "ESRF – Foundation Phase Report." Grenoble

Ewald P. P., 1962: "Fifty Years of X-Ray Diffraction." Oosthoek's Uitgeversmaatschappij, Utrecht, The Netherlands.

Faxén H., 1918: Ann. d. Physik **54**, 615.

Faxén H., 1923: Z. Physik **17**, 266.

Foo E-Ni and Hopfield J. J., 1968: Phys. Rev. **173**, 635.

Freund A., 1971: 2eme Colloque International sur les Methodes Analytiques par Rayonnements X- C.G.R. Toulouse, 129.

Freund A. and Schneider J., 1972: J. of Crystal Growth **13/14**, 247.

Friedrich W., Knipping W. P. and von Laue M., 1912: "Interferenzerscheinungen bei Röntgenstrahlen." in: Sitzungsber. Bay. Akad. der Wissenschaften, Math.-Phys.-Klasse, 303.

Fuji Y., Hastings J. B., Ulc S. L. and Moncton D. E., 1982: SSRL Stanford Report No. VIII-95.

Fukushima K., Onishi T., Shimanouch T., Mizushima S., 1959: Spectrochim. Acta **15**, 236.

Gauss S., 1989: Diplom Thesis, University of Munich.

de Gennes P. G., 1959: Physica **25**, 825.

Gerdau E., Winkler H., Tolksdorf W., Klages C. P. and Hannon J. P., 1985: Phys. Rev. Lett. **54**, 835.

Gehrke R., 1987: private communication.

Gläser W., Egelstaff P. A., 1986: Phys. Rev. A **34**, 2121.

Goulon J., Ellaume P. and Raoux D., 1987: Nucl. Instr. and Meth. **A254**, 192.

Graeff W. and Materlik G., 1982: Nucl. Inst. Meth. **195**, 97.

Grimsditch M. H. and Ramdas A. K., 1975: Phys. Rev. B **11**, 3139.

Gudat W., 1987: in "Synchrotronstrahlung in der Festkörperphysik." 18. IFF-Ferienkurs, Kernforschungsanlage Jülich, Jülich (Summer School proceedings).

Gupta V. D. and Krishnan M. V., 1970: Chem. Phys. Lett., **67**, 165.

Hansen J. P. and McDonald I. R., 1976: "Theory of simple liquids." (Academic New York).

Hastings J. B., Siddons D. P., van Bürck, Hollatz R. and Bergmann U., 1990: Phys. Rev. Lett. submitted.

HASYLAB, 1987: Jahresbericht HASYLAB, DESY, Hamburg (Institutional report).

Haubold H. G., 1975: J. Appl. Cryst. **8**, 175.

Heidemann A., 1970: Z.Phys. **238**, 208.

Hildebrandt G.,1979: Acta Cryst. A **35**, 696.

Hofmann W. G., 1989: Thesis, University of Bayreuth.

Horton G. K. and Maradudin A. A., 1974: "Dynamical Properties of Solids", Vol. 1 (North-Holland, Amsterdam).

Hotz M., 1991: Diplom thesis, in progress

Van Hove L., 1954: Phys. Rev. **95**, 249.

Illini Th., 1991: Thesis, University of Munich.

Illini Th., Burkel E. and Peisl J., 1989: Jahresbericht HASYLAB 1989, Hamburg (Institutional report).

Jackson J. D., 1962: "Classical Electrodynamics" (Wiley, New York).

James R. W., 1948: "Optical Principles of the Diffraction of X-rays" (G. Bell and Sons, London).

James R. W., 1963: *The Dynamical Theory of X-Ray Diffraction* in "Solid State Physics," Vol. **15**. Ed. by Seits F. and Turnbull D., 53, (Academic Press, New York and London).

Jacobsen E. H., 1955: Phys. Rev. **97**, 654.

Joynson R. E., 1954: Phys. Rev. **94**, 851.

Kloos T., 1973: Z. Phys. **265**, 225.

Koch E. E., Eastman D. E., Farge Y., 1983: in "Handbook on Synchrotron Radiation," Vol. **1A**. Ed. by Eastman D. E. and Farge Y. (North-Holland, Amsterdam).

Kohra K. and Matsushita T., 1972: Z. Naturforsch. **271**, 484

Kubo R., 1966: Repts. Progr. Phys. **29**, 255.

Landolt-Börnstein, 1981: New Series III/1. Ed. by K. H. Hellwege (Springer Berlin, Heidelberg, New York).

von Laue M., 1940: "Röntgenstrahlinterferenzen." Akademische Verlagsgesellschaft, Frankfurt a.M..

Laval J., 1938: Acad. Sci., Paris, **207**, 169; **208**, 1512.

Laval J., 1939: Bull. Soc. franc. mineral. **62**, 137.

Levesque D., Velet L., Kürkijarvi J., 1973: Phys. Rev. A **7**, 1690.

Lovesey S. W., 1971: J. Phys. C: Sol. State Phys., **4**, 3057.

Lovesey S. W. and Springer T. (eds.) 1977: "Dynamics of Solids and Liquids by Neutron Scattering." Topics in Current Physics, Vol. **3** (Springer, Berlin, Heidelberg, New York).

Lovesey S. W., 1984: "Theory of Neutron Scattering from Condensed Matter." Vol.1, International Series of Monographs on Physics No. 72 (Clarendon, Oxford).

Lindhard J., 1954: Kgl. Dans. Vidensk. Selsk. Mat. Fys. Medd **28**, 3.

Lyon K. G., Sallinger G. L., Swenson C. A., White J., 1977: J. Appl. Phys. **48**, 865.

Magerl A., Heidemann A., Holm C. and Sirtl E., 1984: Institue Max von Laue - Paule Langevin, Internal Scientific Report No. 84MA24T (Grenoble).

Marsh R. E., 1958: Acta Cryst. **20**, 550.

McSkimin H. J. and Andreatch P., 1972: J. Appl. Phys. **43**, 2944.

Moissev M., Nikitin M., Fedorov N., 1978: Sov. Phys. J. 21, 332.

Moncton D., Hastings J. B., Siddons P. and Brown G., 1986: SSRL Report (Stanford).

106   References

Mountain R. D., 1966: Rev. Mod. Phys. **38**, 205.

Moncton D. E. and Brown G. S., 1983: Nucl. Instr. and Meth. **208**, 579.

Okada Y. and Tokumaru Y., 1984: J. Appl. Phys. **56**(2), 314.

Okasaki A. and Kawaminami M., 1973: Mat. Res. Bull. **8**, 545; Jap. J. Appl. Phys. **12**, 783.

Peierls R. E., 1955: "Quantum Theory of Solids" (Clarendon, Oxford).

Peisl H., Spalt H. and Waidelich W., 1967: Phys. Stat. Sol. **23**, K 75.

Peisl J., 1976: in "Defects and Their Structure in Nonmetallic Solids." Ed. by Henderson B. and Hughes A.E. (Plenum, New York), 381.

Petri E. and Otto A., 1975: Phys. Rev. Lett. **34**, 1283.

Pines D. and Nozières P., 1966: "The Theory of Quantum Liquids." (Benjamin, New York).

Pinsker Z. G., 1978: "Dynamical Scattering of X - Rays in Crystals" (Springer, Berlin - Heidelberg - New York).

Pintschovius L., 1990: Festkörperprobleme **30**, 183.

Platzman P. M. and Tzoar N., 1970: Phys. Rev. B **2**, 9, 3556.

Platzman P. M., 1974: in "Elementary Excitations in Solids, Molecules, and Atoms Part A." Ed. by Devreese J.T., Kunz A.B. and Collins T.C., Nato Advanced Study Institutes Series, Series B: Physics, (Plenum, London, New York).

Platzman P. M. and Eisenberger P., 1974: Phys. Rev. Lett. **33**, 152.

Platzman P. M. and Tzoar N., 1970: Phys. Rev. B **2**, 9, 3556.

Prade J., 1989: Thesis, University of Regensburg.

Rahman A., 1974: Phys. Rev. A **9**, 1667.

Raether H., 1980: in "Excitations of Plasmons and Interband Transititons by Electrons." Springer Tracts Mod. Phys. Vol. **88**. Ed. by Höhler G. and Niekisch E.A. (Springer, Berlin).

Ruppersberg H. and Speicher W., 1976: Z. Naturforsch. **31** a, 47.

Sachs G. and Weerts J., 1930: Z. f. Phys. **60**, 481 and Z. f. Phys. **64**, 344

De Schepper I. M. and Cohen E. G. D., 1980: Phys. Rev. A **22**, 287.

Schrödinger E., 1914: Phys. ZS **15**, 79, 497.

Schülke W. and Lautner W., 1974: Phys. Stat. Solidi (b) **66**, 211.

Schülke W., Nagasawa N. and Mourikis S., 1984: Phys. Rev. Lett. **55**, 2065

Schülke W., Nagasawa N., Mourikis S.and Lanzki P., 1986: Phys. Rev. B **33**, 6744

Schülke W., Bonse U., Nagasawa, Mourikis S. and Kaprolat A., 1987: Phys. Rev. Lett. **59**, 1361

Schwinger J., 1949: Phys. Rev. **75**(12), 1912

Seoud A. E.-H., 1983: Thesis, Universität München

Siddons D. P., Hastings J. B., Moncton D., Hewitt R. and Brown G., 1986: Annual Report, Stanford Synchrotron Research Laboratory.

Siddons D. P., Hastings J. B. and Faigel G., 1988: Nucl. Instr. and Meth. **A266**, 329

Sinn H., Burkel E., 1991: in progress.

Sinn H., 1991: Diplom Thesis, University of Munich.

Simpson H. J. Jr. and Marsh R. E., 1966: Acta. Cryst. **20**, 550.

Sjögren L., 1979: J. Phys. C **12**, 425.

Sjögren L., 1980: Phys. Rev. A **22**, 2883.

Söderström O., Dahlborg U. and Gudowsky W., 1985: J. Phys. F **15**, L23.

Sköld K. and Rowe J. M., 1972: in "Proc. Neutron Inelastic Scattering." IAEA Meeting Paper No. SM-155/D5, Vienna.

Smither R. K., Forster G. A., Bilderback D., Bedzyk M. Finkelstein F., Henderson C., White J., Berman L., Stefan P. and Overluizen T., 1989: Proc. SRI 1988 Conf. Tsukuba, Japan, Rev. Sci. Instr.

Solvia, 1987, 1989: Solvia Engineering AB, Östra Ringvägen, (Västerås, Sweden).

Srivastava R. B. and Gupta V. D., 1972: Indian J. Pure Appl. Physics **10**, 596.

Stedman R., Amilius Z., Pauli R. and Sundin O., 1976: J. Phys. F **6**, 157.

Steyerl A. and Steinhauser K. A., 1979: Z. Physik B **34**, 221.

Strauch D., 1990: private communication.

Strohmeier M., 1988: unpublished report, University of Munich.

Strohmeier M., 1989: Diplom Thesis, University of Munich.

Strohmeier M., Burkel E. and Peisl J., 1989: Jahresbericht HASYLAB, Hamburg (Institutional report).

Stuhrmann H. B., 1980: Acta Cryst. **A36**, 996.

Sturm K., 1982: Adv. Phys. **31**, 1.

Sturm K. and Oliveira L. E., 1984: Phys. Rev. B **30**, 4351.

Sturm K. and Oliveira L. E., 1989: Europhys. Let. **9**, 257; Phys. Rev. B **40**, 3672.

Sturm K., 1990: private communication

Suck J. B. and Gläser W., 1972: in "Proc. Neutron Inelastic Scattering." IAEA Meeting, Vienna, 435.

Sugai S., Ucida H., Takagi H., Kitazawa K. and Tanaka S., 1987: J. Appl. Phys. **26**, L 879.

Sykora B. and Peisl J., 1970: Z. Angewandte Physik **30**, 320.

Trinkaus H., 1971: Z. Naturf. **28a**, 980.

Vradis A. and Priftis G. D., 1985: Phys. Rev. B **32**, 3556

Vuong T. H. H., Tsui D. C., Goldman V. J., Hor P. H., Meng R. L. and Chu C. W., 1987: Solid State Commun. **63**, 525.

Waller I., 1923: Z. Physik **17**, 398.

Walker C. B., 1956: Phys. Rev. **103**, 547.

Wang C. H. and Storms R. D., 1971: J. Chem. Phys. **55**, 3291.

Wang Y. R. and Overhauser A. W., 1988: Phys. Rev. B **38**, 14, 9601.

Warren J. L., Yarnell J. L., Dolling G. and Cowley R. A., 1967: Phys. Rev. **158**(3), 805.

Waseda Y., 1980: "The Structure of Non-crystalline Materials" (McGraw-Hill, New York).

Windisch D. and Becker P., 1990: Phys. Stat. Sol. (a) **118**, 379

Wochner P., Burkel E., Peisl J., Zeyen C. M. E. and Petry W., 1990: Springer Proceedings in Physics, **37**. Ed. by Richter D., Dianoux A. J., Petry W. and Teixeira J. (Springer, Berlin, Heidelberg), 280.

Zacharias P., 1975: J. Phys. F **5**, 645.

Zippelius A. and Götze W., 1978: Phys. Rev. A **17**, 414.

Subject Index

Absorption 16, 49
Acoustic mode 61
Adiabatic approximation 57
Alanine 77
Aluminum 1
Amino acids 77
—alanine 77
—glycine 77
Amorphous sample 43, 69
Analyser 13, 17, 20, 30, 39, 53, 92, 95
Anode material 30, 51
Anomalous scattering 92
Application of inelastic scattering 57
Atomic form factor 11, 16, 57, 92

Backscattering
—geometry 4, 15, 17, 37
—of x–rays 4
—perfect 14
Background 37, 53, 86
Band gap 89
Band structure 87, 89
Bayreuth spectrometer 52
BCS theory 65
Beam divergence 29, 56
Bending magnet 26, 29
Bending modes of molecules 79
Bent crystals 19, 21, 54
—spherically 19, 38, 39
Beryllium 59, 66
—window 36, 49
Biological macromolecules 77
Bose occupation factor 62
Born approximation 8
Bragg law 14
Bremsstrahlung 4, 25
Bridgeman method 86
Brightness 28, 31, 95
Brilliance 28
Brillouin 59
—doublet 71
—scattering 3, 63, 78, 94
—zone, first 68

Calibration factor 43
CCD camera 39
Cesium 72
Ceramic superconductor 65
Characteristic line 4, 51
Charge density 11
Collective
—excitation 9, 66, 84, 91
—density mode 76
—propagating mode 75
Compressibility 73
Compton scattering 9
Conductivity, thermal 37
Constant
—E scan 86
—Q scan 14, 41
Cooling of the monochromator 45
Core level excitation 3
Creation of phonons 58
Cross section
—double differential 88, 92
—intrinsic 10
—Thomson 10
Crystal potential 87
Cutoff wavevector of plasmon 85

Darwin curve 15, 37
D–band metals 89
Debye–Scherrer ring 53, 67
Debye–Waller factor 16, 57
Defects in crystals 91
Degeneration of phonon modes 62
Density of states 70, 71
Detector 33
—arm 34
Diamond 61, 66
—elastic constants 64
—phonon dispersion 61
Dielectric function 10, 83
Diffractometer 34
Diffuse quasielastic scattering 91
Dipole 92
Dispersion of

N. G. Chetaev

Theoretical Mechanics

Translated from the Russian by I. Aleksanova

1989. 407 pp. 190 figs. Hardcover
ISBN 3-540-51379-5

This university-level textbook reflects the extensive teaching experience of
N. G. Chataev, one of the most influential teachers of theoretical mechanics in the Soviet Union. The mathematically rigorous presentation largely follows the traditional approach, supplemented by material not covered in most other books on the subject. To stimulate active learning numerous carefully selected exercises are provided. Attention is drawn to historical pitfalls and errors that have led to physical misconceptions.
Extensive appendices contain material from additional lectures on optics and mechnics analogies, Poincaré's equation and the special theory of elasticity.

Distribution rights for the socialist countries, India and Iran:
V/O "Mezhdunarodnaya Kniga", Moscow

D. Park, Williams College, Williamstown, MA

Classical Dynamics and Its Quantum Analogues

2nd enl. and updated ed. 1990. IX, 333 pp. 101 figs. Hardcover ISBN 3-540-51398-1

The primary purpose of this textbook is to introduce students to the principles of classical dynamics of particles, rigid bodies, and continuous systems while showing their relevance to subjects of contemporary interest. Two of these subjects are quantum mechanics and general relativity. The book shows in many examples the relations between quantum and classical mechanics and uses classical methods to derive most of the observational tests of general relativity. A third area of current interest is in nonlinear systems, and there are discussions of instability and of the geometrical methods used to study chaotic behaviour. In the belief that it is most important at this stage of a student's education to develop clear conceptual understanding, the mathematics is for the most part kept rather simple and traditional.
This book devotes some space to important transitions in dynamics: the development of analytical methods in the 18th century and the invention of quantum mechanics.

A. Hasegawa, AT&T Bell Laboratories, Murray Hill, NJ

Optical Solitons in Fibers

2nd enl. ed. 1990. XII, 79 pp. 25 figs.
Softcover ISBN 3-540-51747-2

Already after six months high demand made a new edition of this textbook necessary. The most recent developments associated with two topical and very important theoretical and practical subjects are combined: **Solitons** as analytical solutions of nonlinear partial differential equations and as lossless signals in dielectric **fibers.** The practical implications point towards technological advances allowing for an economic and undistorted propagation of signals revolutionizing telecommunications. Starting from an elementary level readily accessible to undergraduates, this pioneer in the field provides a clear and up-to-date exposition of the prominent aspects of the theoretical background and most recent experimental results in this new and rapidly evolving branch of science. This well-written book makes not just easy reading for the researcher but also for the interested physicist, mathematician, and engineer. It is well suited for undergraduate or graduate lecture courses.

Springer-Verlag
Berlin
Heidelberg
New York
London
Paris
Tokyo
Hong Kong
Barcelona
Budapest

A. G. Sitenko, Academy of the Ukrainian SSR

Scattering Theory

1991. XI, 294 pp. 32 figs. (Springer Series in Nuclear and Particle Physics)
Hardcover ISBN 3-540-51953-X

This book is an introduction to nonrelativistic scattering theory. The presentation is mathematically rigorous, but is accessible to upper level undergraduates in physics. The relationship between the scattering matrix and physical observables, i. e. transition probabilities, is discussed in detail. Among the emphasized topics are the stationary formulation of the scattering problem, the inverse scattering problem, dispersion relations, three-particle bound states and their scattering, collisions of particles with spin and polarization phenomena. The analytical properties of the scattering matrix are discussed. Problems round off this volume.

B. N. Zakhariev, Moscow; **A. A. Suzko,** Minsk, USSR

Direct and Inverse Problems

Potentials in Quantum Scattering

1990. XIII, 223 pp. 42 figs.
Softcover ISBN 3-540-52484-3

This textbook can almost be viewed as a "how-to" manual for solving quantum inverse problems, that is, for deriving the potential from spectra or scattering data and also, as somewhat of a quantum "picture book" which should enhance the reader's quantum intuition. The formal exposition of inverse methods is paralleled by a discussion of the direct problem. Differential and finite-difference equations are presented side by side. The common features and (dis)advantages of a variety of solution methods are analyzed. To foster a better understanding, the physical meaning of the mathematical quantities are discussed explicitly. Wave confinement in continuum bound states, resonance and collective tunneling, energy shifts and the spectral and phase equivalence of various interactions are some of the physical problems covered.

R. M. Dreizler, University of Frankfurt; **E. K. U. Gross,** University of Würzburg

Density Functional Theory

An Approach to the Quantum Many-Body Problem

1990. XI, 302 pp. 18 figs. Hardcover
ISBN 3-540-51993-9

Density Functional Theory is a rapidly developing branch of many-particle physics that has found applications in atomic, molecular, solid state and nuclear physics. This book describes the conceptual framework of density functional theory and discusses in detail the derivation of explicit functionals from first principles as well as their application to Coulomb systems. Both non-relativistic and relativistic systems are treated. The connection of density functional theory with other many-body methods is highlighted. The presentation is self-contained; the book is thus well suited for a graduate course on density functional theory.

Springer-Verlag
Berlin
Heidelberg
New York
London
Paris
Tokyo
Hong Kong
Barcelona
Budapest

Springer Tracts in Modern Physics

* denotes a volume which contains a Classified Index starting from Volume 36